Mastering Torch-Fired Enamel Jewelry

the next steps in
painting *with* fire

BARBARA LEWIS

NORTH LIGHT BOOKS
Cincinnati, Ohio
CreateMixedMedia.com

Contents

For bonus projects, tips, techniques and more visit:

CreateMixedMedia.com/Mastering-Torch-Fire

Tools & Materials Used in This Book

You don't have to acquire all of these tools and materials at once. Begin with the tools and materials discussed in the Getting Started section and add a few of the tools and materials required for one or two of the projects. As you progress through the book, add more tools to continue building your tool kit.

BASIC TORCH-FIRED ENAMELING KIT

- angle bracket
- Bead Pulling Station
- bread pan
- c-clamp or easy release clamp
- fuel (MAP gas, butane or propane)
- heat-resistant surface
- hose clamp
- mandrels
- Oil-Dri, vermiculite or kitty litter
- torch (Hot Head)
- water jar
- work table

ENAMELS

- variety of enamels (80 mesh, 6/20, liquid and liquid dry form, overglazes and supplements) as per project material lists

METALS USED IN THIS BOOK

- copper screen, 80 mesh
- copper sheet, 24-gauge
- copper wire
- rebar wire, 16-gauge
- sterling silver wire, 16- and 24-gauge

BEADS USED IN THIS BOOK

- bead caps
- brass: angel wings, discs, flowers
- copper: beads, discs, gears
- heisi beads
- pearls
- pierced temple filigree beads
- polymer clay beads

ADDITIONAL JEWELRY SUPPLIES & MATERIALS

- bead-reamer
- bead stoppers
- bead-stringing wire, 19-strand
- beading tweezers
- C-Lon cording
- clamshells
- copper eyelets
- crimp beads
- Eugenia Chan three-hole punch
- leather cording
- rubber earring stop
- stainless-steel cable choker
- variety of clasps (ball-and-hitch, hook-and-eye, magnetic, ball-and-chain, etc.)
- variety of head pins
- variety of jump rings
- variety of metal chains

TOOLS

- bench blocks: rubber and steel
- bench pin
- Crafted Findings Riveting Tool
- dapping block and punches
- drill: Dremel, flexible shaft, household
- files: diamond, metal, needle
- hammers: chasing, ball-peen, riveting, brass mallet
- hole punch
- jeweler's saw and blade
- metal shears
- nail set
- pliers: chain-nose, cross-locking, flat-jawed welding, flat-nose, round-nose
- scissors
- tube wringer
- vise
- wire cutters
- X-ACTO knife

SOLDERING TOOLS & MATERIALS

- butane torch
- fire brick
- flux (paste or liquid, hard and soft)
- frit tray
- liquid dish soap (such as Dawn)
- scrubbing pad
- sheet solder
- solder (easy, lead-free stained glass, hard)
- solder brush, liner brush or inexpensive natural bristle brush
- tweezers or soldering pick

OTHER

- adhesive-backed vinyl
- baking soda
- buttonhole cutter
- camera
- ceramic decals
- cotton-filled cording
- Crock-Pot
- darkening agent (such as Black Max)
- die-cut machine
- duct tape
- eyedropper
- G bass-guitar string
- liver of sulfur (LOS)
- markers
- mineral spirits
- molding compound
- patina solution (such as Nova-can)
- paper plate
- pickle solution
- pickle pot (Crock-Pot)
- plexiglass
- purse strap
- quick-set glue
- ruler
- paper
- sanding pads
- scrap leather
- sewing machine
- small containers
- T-pins
- upholstery-weight thread
- variety of small paintbrushes
- wax paper

EFFERVESCENT REFLECTION

Effervescent Reflection is a variation on the **Tide Pool** project. For a downloadable how-to PDF, visit createmixedmedia.com/mastering-torch-fire.

Introduction

So you've started enameling but want to take your work to the next level. You want to learn some easy techniques that will brand your work and make it identifiably yours. This book will not only spark creative ideas but will provide you with the tools and resources needed to execute your designs. *Torch-Fired Enamel Jewelry: a Workshop in Painting with Fire* may have been your jumping off point, but now you're thirsting for more.

First, we'll start with an overview of the immersion process of torch-fired enameling. Even though this may be a refresher for you, this groundbreaking enameling technique deserves a second look. Your time in the studio will be more about fun and less about cleaning metal. For lampworkers, this technique will be familiar territory because we'll be using a mandrel—a stainless steel rod—onto which a bead or pendant is placed. In just an afternoon and even with no enameling experience, you'll have created lots of beads and pendants that will become your canvases for further surface decoration.

By surface decoration I mean commercial or home-made decals, custom-tinted liquid enamels, firescale decoration, stencils, controlled overfiring, 6/20 enamel and more. Are you interested in narrative jewelry that allows your work to speak volumes? Decals can help you do that, in the literal sense.

As a Thompson Enamel distributor, I'm surrounded by a hundred different enamel colors, which puts me in a unique position to offer practical advice about how enamels look and behave. Did you know that Pumpkin and Buttercup need more layers, whereas Black will give good coverage with only one layer? Did you know that Quill White is sensitive to carbon trapping? If you like Khaki, this may be a good choice. I will also share favorite color blends developed in the Painting with Fire Studio. As we go through the projects in the book, I'll share these "trade secrets" with you.

We'll fabricate our jewelry using simple and time-saving tools that give professional results.

Because jewelry design, like other things in our culture, responds to economic realities, we'll explore the option of replacing more costly metals with less costly ribbons, leather, cording and fiber strands. These materials will also allow us to weave more color into our work.

So buckle your seatbelt and let's get started!

INNOCENCE
Innocence is a variation on the **Forest Nymph** project.
For a downloadable PDF, visit
createmixedmedia.com/mastering-torch-fire.

Getting Started

It's been three years since *Torch-Fired Enamel Jewelry: a Workshop in Painting with Fire* was released and greeted by an enthusiastic worldwide community of jewelry artists looking for a new, easy and fun way to personalize jewelry through the use of color. That book, which introduced the Immersion Process of torch-fired enameling, generated more good things than I could have envisioned, including being named Best Craft Book of 2011 at Amazon. In addition, the United States Patent and Trademark Office considered the technique so unique that our application for Utility Patent #8,470,399 was granted for the Bead Pulling Station and the enameling process associated with it. The Immersion Process relies heavily on the functions of the Bead Pulling Station for safety and speed.

But let's say you're coming to the party late and you've never heard the word enamel applied to anything other than spray paint or the outer layer of your teeth! If this is you, you're in for a wake-up call and a whole bunch of fun! As it relates to jewelry, lustrous copper vessels or exquisite Faberge eggs, enamel is defined as *glass on metal*. If we put the word *vitreous* before the word enamel, we have clarified that we're talking about glass. This is a very important distinction. Because of the popularity of the art of enameling, and especially torch-fired enameling, more and more products labeled as *enamel* are being introduced to the marketplace.

Historically, enameling required a kiln to fuse powdered glass to metal, as well as a good deal of education about the process. But then some savvy folks started using inexpensive torches as the "heat" part of the enameling process. Even so, the fussy steps of the enameling process remained, whether firing in a kiln or torch firing using a tripod. Scrubbing the metal to a pristine condition was still required so oils (from hands) or environmental grime did not present a barrier to enamel adhesion.

Adhesives were required to get the enamel to stick to the metal. Counter enameling, where enamel is applied to the back of the piece being enameled to prevent warping, is yet another step. Even when the artist could indulge in the expense of a kiln, these enameling steps had many of us shying away from trying it. The good news is that NONE of these steps is a part of the Immersion Process of enameling.

The success of this technique is due to a single factor: We heat the metal first. This single step sets the Immersion Process apart from kiln firing and every other type of torch-firing method. No tedious cleaning of the metal in necessary because the flame cleans the metal. And the metal is hot, so there's no need for an adhesive because enamel sticks to the hot metal. There is also no need for a separate counter-enameling step. When we dredge the metal through the enamel, we cover the front and the back at the same time.

In addition to the time-saving features of the Immersion Process, there are also some added safety features. Fewer glass particles become airborne because we don't sift the enamel onto the metal. Our torch is also firmly attached to our work surface, allowing both hands to be free. It also means we are well aware of the direction of the torch flame. In a workshop situation this advantage can be huge!

Have I mentioned speed? Using the Immersion Process and the Bead Pulling Station, we can fire an iron bead with three layers of enamel in less than a minute! Yes, you heard me right, I said *iron*. My first book introduced inexpensive iron filigree beads to the enameling community. Can you think of a better way to add value to your jewelry without an equivalent cost? While this book has been in the creation stage, we've seen sterling silver skyrocket to $40 a troy ounce before leveling out at $23. In 2000, it was less than $5 a troy ounce!

Through diligent work, the team at Painting with Fire has identified a type of brass that accepts enamel. Juicy transparent enamels that pool in the recesses of ever-popular brass stampings help to accentuate the design.

The Immersion Process of enameling involves placing a bead, pendant or charm on a stainless steel rod (mandrel), which does not conduct heat. We hold the bead in the oxygenated sweet spot of the flame until it glows orange. Then the bead is dredged through a container of enamel. Repeat this process one or two more times, and you're ready to pull the bead off the mandrel using the Bead Pulling Station. The hot bead falls into a bread pan cushioned with ceramic fiber, vermiculite, kitty litter or Oil-Dri.

But before we go any further into the firing process, let's talk about safety.

Safety

Working with fire and powdered glass has some inherent dangers. Let's see how we can protect ourselves and still have a good time.

VENTILATION

Good ventilation is a must! You need an exchange of air … Out with the old, in with the new! But it needn't be high tech. A room with cross ventilation provided by a wall of windows opposite two open doors was perhaps my best former teaching location outside my own studio. The flameworking studio at the Painting with Fire Studio is a long, narrow room. The room is perfect from the standpoint that all students sit with their torches facing the wall.

However, obtaining adequate ventilation proved to be a challenge. A state-of-the-art-ventilation system was designed so that each student has an exhaust vent that pulls 300 cubic feet per minute away from the torch. Internet research via lampworking sites provided a wealth of information that was synthesized to create an ideal experience in our studio. Be sure to conduct your own research before you begin torch firing.

BURNS

Burns in the studio are usually minimal. More students burn themselves touching a hot bead than any other way. It happens so simply. You've got a bunch of beautiful beads in a bread pan that you can't resist looking at and you think you remember which bead went in last. Ouch! Such burns are not serious, but it does teach you a lesson. Wearing form-fitting cotton clothing is also recommended when working with an open flame, and keep long hair tied back. Singed bangs aren't fun either!

EYE PROTECTION

Clear safety glasses are a good choice. However, if you wear eyeglasses, they may be sufficient protection. Sunglasses can reduce light intensity. But stay away from welding glasses. Even the lowest shade prevents you from seeing the sweet spot of the flame.

Speaking of seeing the flame, too much natural light is your enemy. You will NOT be able to see the flame if firing in natural light. Trying to find the sweet spot of the flame will be impossible. If you must fire outdoors, set up an area where the flame is shielded from sunlight.

MASK

Since we're not sifting enamel, airborne glass particles are greatly reduced. If you want to be extra careful, a 3M Particulate Mask 8233 is a very comfortable and inexpensive mask. Remember to wipe down your work area with a damp cloth and/or wet mop to avoid stirring up enamel particles.

EATING & DRINKING IN STUDIO

We also recommend that you do not eat or drink in the studio near your work surface. This is recommended for all fine arts studios, not only enameling studios. Just use common sense and you'll be fine.

FATIGUE

One final safety tip: Be aware of your fatigue level. Torch firing can be mesmerizing. When you have mental fatigue or your hands are tiring, there's a greater chance for getting a burn or simply producing a lower-quality enameled bead. Just turn off the torch and take a break! Sometimes your break needs to last only a few minutes before you're ready to resume firing.

Workstation and Setup

Now that you have the broader picture, let's get specific. The workstation is designed for safety, efficiency and ease. The Immersion Process is also a low-tech, low-cost method with small space requirements. If you have an extra square foot of space, you can set up a workstation. The components include a heat-resistant surface, a torch with bottle gas hardware, fuel, a jar of water, mandrels, the patented Bead Pulling Station, disposable bread pan, Oil-Dri (or vermiculite or ceramic fiber) and enamels.

HEAT-RESISTANT SURFACE

The heat-resistant surface will protect your countertop from a stray hot bead. It can be an aluminum baking sheet, a ceramic tile or cement board. Since we're not directing the flame at our heat-resistant surface, even cardboard covered in heavy-duty aluminum foil could suffice. Because the Bead Pulling Station slides onto the back of your heat-resistant surface, it should be no thicker than ⅜" (9mm).

FUEL

MAP gas is an ideal fuel and will easily give you great enamel color. However, propane is about one-third the cost of MAP gas. If you choose propane as your fuel source, just hold your metal slightly beyond the sweet spot of the flame when firing. (Refer to *The Immersion Process*, page 20, for more details.)

Butane fuel certainly gets hot enough to fuse enamel. However, in the Immersion Process, when possible, we want to heat our entire metal piece at one time. The Immersion Process requires a bushy flame, which would eliminate a butane torch from being practically used. The cost of fuel needed to heat a 1¼" (3cm) square of copper sheet to a glowing orange heat would quickly make this an expensive choice, not to mention the man hours necessary to keep refilling the torch. A butane torch may be good for trinkets or a few small items in a pinch.

Acetylene is a dirty fuel and using it will affect your enamel colors. As a result, it is not recommended for our work. If you have a dual fuel torch that adds oxygen,

it's worth a try, but remember to be judicious because the addition of oxygen creates a very hot flame.

TORCHES

A Hot Head torch fired all of the projects in this book and is my companion when I teach. To the newbie, the Hot Head may look like any other brass torch sold at the hardware store. However, when you look more closely, you'll find that holes have been drilled into the base of the torch nozzle. These holes allow more oxygen to mix with the fuel, which is crucial to developing bright, clear enamel colors.

If you have a hardware-style torch, you can still get bright colors, but you will need to hold your metal much further out in the flame, well beyond the sweet spot.

The Hot Head torch has a bushy flame that allows for the even heating of larger beads and pendants.

It also produces a very hot flame that allows metal coverage with two layers of enamel instead of three.

Since some of you may have been at this for a while, you may be asking whether it's time to upgrade your equipment. When you find yourself going through MAP gas tanks rapidly, you're probably ready for a dual fuel (propane/oxygen) lampworking torch. I teach with the Hot Head because it's a reliable and safe entry-level torch. However, when I'm working in my studio, my lampworking torch is my preference. I chose an oxygen concentrator instead of an oxygen tank. The concentrator, which produces oxygen on demand, comes in different sizes, based on your needs. I find that all I need is the smallest concentrator, which

produces 5 psi (pounds per square inch). The propane source is a 20-pound tank that we usually associate with our gas grill.

This setup reduces the firing time, gives consistently good enamel color and enables us to fire larger pieces. Another advantage is that I don't worry about low levels of gas in my MAP gas tank darkening my enamel color. There is a greater initial investment of about $500 to $600 (lampworking torch, oxygen concentrator and propane gauge with safety valve). By way of comparison, the Hot Head torch with bottle gas hardware and an initial 1-pound tank of MAP gas costs about $60. I recommend you start with a Hot Head torch and grow into a lampworking setup.

GAS BOTTLE HARDWARE

If you're using a 1-pound fuel tank, you'll need a hose clamp, angle bracket and C-clamp (or easy-release clamp) to attach the torch and heat-resistant surface to the table. If you're using a Hot Head torch, you will want to adjust the angle of the bracket slightly (about 30° upward.) This slight adjustment will change the torch angle to allow more space for your hands to work under the torch.

Figure 1

Figure 2

Figure 3

To make the adjustment, place the bracket in a vise so that the tops of the jaws of the vise are about 1½" from the bend in the bracket. The other leg of the bracket should be facing away from you and will be parallel to the floor. Grab the end of the bracket and give it a pull toward you (see Figure 1). This will place a crimp in the one leg of the bracket. Remove the bracket from the vise. If you don't have access to a vise, you can attach the L-bracket securely to the MAP gas tank and the table, then pull slowly upward on the MAP gas until you get the desired angle. There is a video con my website (paintingwithfirestudio.com) that shows how to make this adjustment.

Now that we have the angle bracket adjusted, set it aside for a moment so you can screw your torch head onto your gas tank (see Figure 2). Place the straight leg of the angle bracket next to the gas tank, about two inches down from the shoulder of the gas tank. Make sure that the extending leg of the bracket is facing in the same direction as the torch head. Starting at the bottom of the tank, slide the hose clamp over the tank and the angle bracket. Tighten the screw on the hose clamp with a flat-head screwdriver (see Figure 3).

BEAD PULLING STATION

The BPS, the workhorse of the Immersion Process, is designed to safely and quickly remove hot enameled objects from the mandrel. The BPS can easily slip onto a heat-resistant surface. Just slide your surface into the channel created by the bend in the BPS. This tool is so unique the United States Patent and Trademark Office issued Utility Patent #8,470,399 to protect the BPS and the immersion method of enameling associated with it. The primary objective for obtaining this patent was to protect the educational integrity of this unique process. The BPS can be used to create beads for your personal use and for sale but cannot be used for fee-based teaching except by licensed and certified teachers of the Painting with Fire process.

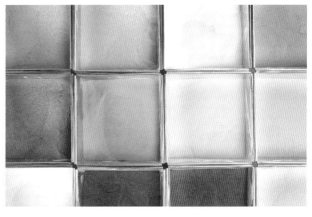

BREAD PAN WITH OIL-DRI, VERMICULITE OR KITTY LITTER

A disposable bread pan filled with Oil-Dri, vermiculite or kitty litter provides a nonflammable cushion onto which your beads can fall.

ENAMELS

Let's not forget our enamels, which you'll have in Pyrex or tin containers. Since I sometimes travel to teach, I need my containers to have tight-fitting lids. However, when I first started, I recycled my tuna and cat food tins. A tray of Pyrex custard cups can be covered by another tray in order to keep the enamel clean. (Plenty more about enamels soon.)

BEADS

A stash of beads in a small container is convenient to have on your workstation.

MANDRELS

Mandrels are stainless steel rods onto which we place the bead for firing. Even though the rod will be in a flame of over 2,000°F, it will stay cool in your hand because stainless steel does not conduct heat. Mandrels come in various diameters and can be purchased at a welding shop or online. Common sizes used in our studio are 1mm, 1.2mm, 2.2mm and 3.5mm in diameter. Nine inches is an adequate length for a mandrel.

WATER JAR

A jar of water allows you to quench your mandrel after firing each bead.

PUTTING IT ALL TOGETHER

Now that we have all of our materials, let's set up our workstation. Place the heat-resistant surface on the table. If you're right-handed, clamp the torch slightly off center to the left. If you are left-handed, clamp the torch slightly to the right of center. Slip the BPS onto the back edge of the heat resistant surface. Place the jar of water to your right, your enamels and beads to the left. Your matches or striker can be placed at your convenience. (The photos here show the workstation set up for a right-handed person. If you're left-handed, just reverse everything!)

The work station setup is designed for safety, efficiency and speed.

Metals

So now that we're all set up, you're probably asking yourself, "What metals can I enamel?" Conventional wisdom dictates that copper, silver and gold are good choices. As a result of their historical use in the field of enameling, you'll find lots of good information on working with these metals. The enameling of iron beads and components, a happy accident that occurred in my studio, was introduced in my earlier book. As a way to add value to our work—while incurring very little expense—the enameling of iron beads has taken the torch-fired enameling community by storm.

Because the hook for me in studio work is the playtime during which unexpected results occur, I set my sights on trying to figure out why some brass pieces enameled while others would not. These inconsistent results kept me entangled in hot pursuit of The Answer. I began to analyze my results and the results of others. The seemingly endless availability of various types of brass complicated the matter.

Brass, actually an alloy of copper and zinc, goes by many different names depending on the ratio of zinc to copper (which can be between 5 and 50 percent). Muntz brass, red brass, yellow brass … it's all very confusing.

But let's back up a bit. Why are we attracted to brass anyway?

Primarily we're attracted to brass for its gold color, its ability to react to patinas and the availability of a wide variety of stampings. With the addition of zinc, a copper alloy is created that is less expensive than solid copper, stronger than solid copper and less susceptible to corrosion. As a jewelry artist shopping at my neighborhood bead store, I usually don't know the zinc/copper ratio of the brass I'm holding in my hands, and, frankly, it tends to make little difference unless I'm going to enamel it. The use of a magnet to detect ferrous metals that would successfully accept enamel is of no use when trying to determine the suitability of brass because brass is not attracted to a magnet.

So, what's the big deal about the zinc content? The brass stamping in my left hand looks the same as the piece in my right, so why won't they enamel the same way? The trouble arises because zinc and enamel have different expansion and contraction rates. If the zinc content is too high, enamel immediately begins to pop off the metal. Thompson Enamel recommends that zinc content be kept to 5 percent or less. This "gilding metal," as it is known, is more readily available in sheet form, but, even then it is not a common item in jewelry-supply catalogs. After years of trial and disappointment, I'm happy to report that Rich Low Brass, with a content of 85 percent copper and 15 percent zinc, yielded the most successful enameling results. The brass must be raw (unplated).

An excellent article on suitable metals appears in the *Thompson Enamel Workbook*, authored by Woodrow W. Carpenter, Bill Helwig and Tom Ellis. It should be noted, though, that since the article was written, Thompson has ceased production of certain enamels, such as those for aluminum. As you can see, enamel availability is another factor in determining which metals can be enameled.

Melting temperature is also a major consideration. Pewter melts below the maturing temperature of medium temperature/medium expansion enamels, so it cannot be used.

Enamels and Supplements

So now that we know more about which metals we can use, what do we need to know about enamel? First, keep in mind that the definition of enamel is glass on metal. Enamel is ground glass, which is then further defined by mesh size, melting temperature, expansion rate, color, transparency and opacity. The most commonly used enamel is formulated for copper, silver and gold … but now we know it works on brass and iron, too.

All About Enamels

MESH SIZE

Eighty-mesh enamel is the most commonly used enamel size. Eighty mesh means that in a linear inch of screen with eighty openings all of the enamel will sift through. There are other mesh sizes of enamel, which we'll discuss a little later.

EXPANSION AND TEMPERATURE

Medium temperature/medium expansion indicates that the enamel matures (becomes shiny) at temperatures between 1,400°F and 1,500°F and is formulated to have the same expansion rate as copper, silver and gold. When enamel chips from the surface of the metal (as in our experiments with brass), it usually is because there is an incompatibility in the expansion rates.

COLOR

We can all agree that an allure of enamel is its color. With more than 150 individual enamel colors and variations derived from layering one enamel over another, we can color match nearly any available fabric color. But not all enamels are created equal, especially when it comes to price. Variations can be attributed, many times, to the oxides used to produce a particular color. An enamel price could be affected by the rarity, value and location of an ingredient. For instance, purples and pinks contain gold (the metal), and cobalt is mined in the Democratic Republic of the Congo and in Zambia. The destabilization of a government and the commodities costs of other rare metals affects the price of the enamels we use.

TRANSPARENCY AND OPACITY

Transparent enamel is great for accentuating details because the enamel flows and fills the depressions of the design. The color of a transparent enamel is very much affected by the color beneath it, be it a dark metal or another enamel color. As a result, the best way to blend color is by layering transparent enamels, either over other transparent enamels or over opaque enamels. Transparent enamels are shinier. Clear transparent enamel, which also is called flux, can be layered over opaque enamel to create a glossier surface.

Opaque enamels give good coverage and cannot typically be layered over one another to create another color, because the top enamel will hide the color of enamel beneath it. You cannot see through opaque enamels. Opaque enamels will also obscure the details of a metal piece, which is why many of us prefer to use transparent enamels over brass stampings and filigree beads

TORCH-FIRED TIP

Opaque enamels begin with the number one, as in 1870 Orient Red. Just remember, the words *opaque* and *one* both begin with the same letter *O*! So, without any other identifying information, I know that 1870 is opaque enamel. On the other hand, transparent enamels start with the number two, as in 2305 Nile Green. The word *transparent* and the word *two* both start with the letter *T*. To take it one step further, enamels that begin with the number *one* go on *first*, enamels that begin with the number *two* go on *second* (or last).

Popular Enamel Supplements

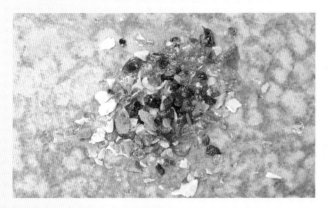

6/20 ENAMEL

6/20 mesh refers to the particle size. It means that in a linear inch of screen with six openings, all of the enamel will sift through, but in a linear inch of screen with twenty openings, all of the enamel particles will lie on top of the screen. 6/20 mesh is a common size for glass bead making and our favorite size for making enamel head pins. You can also lay this type of enamel on top of a pendant and create either raised bumps or liquid pools of glass. If you want your enamel to remain raised, heat from underneath just long enough for the enamel to become fused to the metal (or the base enamel), then bring the torch topside to round off the corners and edges of the enamel.

LIQUID ENAMEL

Liquid enamel is a favorite, too! In my studio, I primarily use the dry form, BC1070 White or BC303L Clear. Liquid enamel also comes in a wet form, but I don't like to pay to have water shipped (it's heavy). Dry form liquid enamel, which resembles confectioners' sugar in its fineness, contains binders that help to keep the particles in suspension when water is added.

ENAMEL THREADS AND CAT WHISKERS

Enamel threads are about 2mm in diameter and about three inches in length. Cat Whiskers are thinner and shorter. If you lay a thread or whisker on top of your enamel and heat from below, the thread or whisker will melt as a straight line. If you direct your flame topside, the thread will ball up.

MILLEFIORI

I love millefiori, which means "thousand flowers" in Italian. It comes in different sizes, but the tiny glass wafers, which are about 3mm in diameter, melt very easily with a torch. When heated, the wafers can expand in size to about three-eighths of an inch.

P-1 OVERGLAZE ENAMEL AND P-3 UNDERGLAZE ENAMEL

These enamels, which are creamy and easy to paint with, are pitch black. After you paint your design, you have to refire the piece. P-1 does not require a coat of clear enamel; P-3 does.

The Immersion Process

In the Introduction, we celebrated the advantages of the Immersion Process of torch-fired enameling. Now it's time to get started! We've got our workstation set up for efficiency and safety. We're in a space with good ventilation, dimmed lighting, and maybe some Aerosmith or Andrea Bocelli playing in the background.

And now a bit of advice. If you follow these three easy steps every time, you'll have beautifully enameled pieces!

1 Reduce the gas after getting the torch started. This will create a more balanced mixture of air and fuel, and lead to gorgeous enamel work and gas savings.

2 Fire the bead in the sweet spot of the flame. You may think that the flame nearest the torch nozzle is the hottest part of the flame. In fact, it is the coolest part of the flame and rich with fuel. This excess fuel will create muddy-colored beads.

3 Keep the bead ¼" (6mm) to ⅜" (9mm) inch from the end of the mandrel in order to prevent a buildup of enamel on the mandrel. A buildup can also close the hole of the bead when you're pulling the bead off the mandrel.

Okay, now that you know the secrets to success, I'll fill in the blanks.

Sit straddling the torch. This is a much more comfortable position than sitting sidesaddle and twisting your body to reach the flame.

Light your match and turn the knob on the torch to the left until you hear the gas escaping. Bring the lit match up from beneath the very end of the torch nozzle. Have the flame of your match touching the torch nozzle. If you place your lit match too far in front of the torch nozzle, the air and gas will blow out the flame of the match. If the torch doesn't light, try giving it a little more gas.

Now that your torch is lit, reduce the gas from a roar to a more gentle whooshing sound. This will help to properly balance the amounts of air and fuel. If you fail to reduce the flame, you'll be firing in a fuel-rich environment that will create muddy beads. Lowering the flame will also save gas and money!

1 Slide a metal bead onto a mandrel and place the bead in the sweet spot of the flame, which is where you'll find the ideal mixture of gas and oxygen. This is the hottest part of the flame and the spot that will give you the best enamel color. With MAP gas, the sweet spot is just at the tip of the inner blue cone of the flame, or about 2½" to 3" (6cm to 8cm) from the end of the torch nozzle. If firing with propane, it's a tad beyond that point.

If you are firing a round bead, quickly spin the mandrel in your fingers to evenly heat the bead. When it glows orange, it's time to dredge the bead through the enamel.

2 Tilt the container of enamel toward the bead. Don't just tickle the enamel by spinning the bead on the surface. Really dig the bead into the enamel. Make sure the tip of the mandrel touches the bottom of the container and pull the mandrel through the enamel. This technique will ensure that both ends of the bead get covered.

3 Tap the mandrel on the edge of your container to knock off any excess enamel before moving the bead back into the flame. Fire until the bead turns orange.

4 You may have noticed that the bead has been pushed further away from the end of the mandrel. This is where the Bead Pulling Station comes into play. Reheat the bead slightly and then rest your mandrel in one of the *V* notches on the BPS. With the bead in the interior space, pull your hand toward your body. This will pull the bead down toward the end of the mandrel until it is back at the ideal spot, ¼" to ⅜ " (6mm to 9mm) from the end.

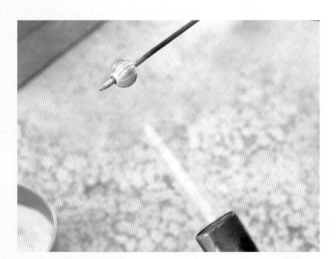

5 Continue firing. Repeat the heating and dredging process for a total of two to three layers of enamel if you are using the Hot Head torch. (You may need more layers if using a different torch, such as a hardware store torch or a Fireworks torch.)

6 Remove the bead from the mandrel using the Bead Pulling Station. The goal here is to pull off the bead swiftly so that it is just dangling on the end of the mandrel. The end of the mandrel should be in the interior space of the bead.

If you start to feel resistance when you're pulling off the bead, it's fair to assume that some of the enamel has cooled and hardened the bead onto the mandrel. To continue to pull the bead will create a problem. Look at the end of the bead hole closest to your hand to see if some of the enamel is being pulled off the bead. If so, direct the flame at this point to melt the connection between the enamel and the mandrel. Now remove the bead. If the bead still resists, you may have to lightly reheat the entire bead before pulling it off in the BPS.

TORCH-FIRED TIPS

- Never walk away from a lit torch. Never leave the gas escaping from the tank without the torch being lit. MAP gas is heavier than air and will not dissipate in it. The gas could settle near your torch and be ignited by a spark. Use common sense and you'll be fine.

- Choose a mandrel that is as close to the size of the hole in your bead as possible. Place the bead on the mandrel so that it is ¼" to ⅜" (6mm to 9mm) from the end of the mandrel. Hold the container of enamel in your non-dominant hand.

- It is important to hold the container of enamel in your hand while firing. You need to know where it is because valuable seconds can be lost while you're trying to find the right container of enamel. The metal heats up fast, but it also cools down fast. Enamel will not fuse to cool metal.

- A foolproof way to find the sweet spot is to hold a mandrel horizontal to the table and perpendicular to the flame. Insert the tip of the mandrel into the flame starting at the torch nozzle. Slowly move the mandrel further out in the flame. When the mandrel glows brightly, you've found the sweet spot!

- All of your movements will be small and fast. If you get the bead glowing, don't waste that advantage by being slow to get it into the enamel. Get the bead in fast!

- If you've fired consistently for several hours, you'll notice that your gas tank may start to freeze up. The bottom of the tank is cold and condensation may appear on the outside of the tank. You can unclamp the torch and put the tank, for a minute or two, in a container of warm water. Be careful not to burn yourself on the hot torch. Tank freezing is an indication that you're getting low on gas, but you'll probably have an hour or two of firing time left. As your tank gets low, you also may notice that your enamel colors may become a bit muddy.

- The key to an even application of enamel is to have the bead evenly heated and to quickly rotate the bead in the enamel.

- To remove excess enamel on the end of the mandrel, allow the enamel to cool and squeeze with pliers to crack it off.

- For large beads, it may take several pulls of the bead at the BPS to remove. Slightly reheat the bead between pulls if necessary.

- Beginners have a tendency to overheat the bead because it's hard to believe the technique is this easy. It is this easy. Believe it and trust in it.

Firing Flat Objects

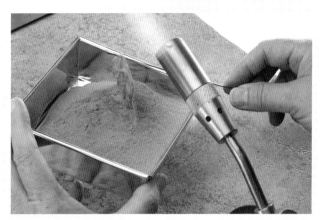

1 Make the hole in your piece slightly larger than normal because enamel will definitely narrow the hole. Place the piece ½" (13mm) from the end of the mandrel. You need this extra allowance so the pendant doesn't fall off during firing or dredging. Allow the pendant or charm to dangle in the flame. It will probably take longer to heat than a round bead.

2 When the piece glows, dredge it through enamel. Make sure you completely immerse the piece in enamel. This technique eliminates a separate counter-enameling step.

3 Fire the piece until it glows, and then dredge it again. Pendants, especially those made of copper, may take four or five coats of enamel to get them adequately covered because they're not usually as hot as filigree iron beads when they enter the enamel.

4 Remove the piece from the mandrel. You're going to have to remove pendants and charms from the mandrel in a slightly different manner. Because enamel will build up on the end of the mandrel, you can't pull the pendant off without the excess enamel filling the hole. See *Alternate Ways to Remove a Pendant*, page 34, for more information.

Firing Flat Objects Without a Hole

If you're firing a metal piece that does not have a hole, simply use pliers to grasp onto an edge of the piece. Enamel the section of metal opposite the pliers. Work your way around the piece, gripping the enamel that has cooled with pliers.

Another alternative would be to use the torching basket created by Eugenia Chan. The use of the torching basket is similar to that of firing a piece of metal placed in a trivet on top of a tripod.

There is one major exception, however. The torching basket has a very open design with much less metal to act as a heat sink. A heat sink absorbs much of the heat meant to be delivered to the object being heated. With the torching basket, the metal heats quickly, which means there is less chance of over-firing the enamel. A real advantage for those who use the Immersion Process of enameling is that the handle of the torching basket allows the piece to be held in the flame without any necessary adjustments to our torch-firing setup. However, you can always use the torching basket with a tripod while still experiencing the benefits of quick and easy heating. In this case, the metal would be stationary and the handheld torch would circle the metal for even heating.

Firing Delicate Shapes

Some metal pieces, like bead cones, are so delicate that if you were to try to heat the piece in its entirety, you would surely burn a hole in part of the metal. Instead heat the piece a section at a time. Hold the container of enamel below the end of the torch. Heat a section, then dip. These are very small and fast movements. Continue around the piece until it is entirely covered in enamel. Because glass is an insulator, once you get enamel around the entire piece, you can heat it more aggressively. Continue to dredge through the enamel until you're satisfied with the coverage. I usually apply an extra layer of enamel to the bottom edge of such pieces.

Refiring Enamel

One of the coolest things about enameling? You get many chances to get it right! The main consideration in refiring is that you allow ample time for the piece to warm up slowly in the flame. If you rush it, you risk thermal shock. Enamel will pop off the metal or pull away in sections leaving an ugly mess.

Simply warm the bead or pendant in the nonvisible part of the flame (7"–10" [18cm–25cm] from the end of the torch). When it starts to darken in color, bring the piece gradually closer to the sweet spot. When the piece glows orange, dredge it through enamel!

Firing Decals

The decals we use on torch-fired enamel are called *ceramic waterslide decals*.

Homemade decals can be printed on older-model laser printers that use ink that contains high percentages of red iron oxide. Red iron oxide is a colorant that survives the firing. Because there is a lot of interest in decals for ceramics, glass and enamel, you'll find lots of information about homemade decals on the Internet. Since it's not possible to provide a listing of printers that will or will not work, do some Internet research.

Commercial ceramic decals also can be purchased online. The images are created with pigments that survive the maturing temperature of our enamel, which is 1,400°F–1,500°F. These decals are widely used to decorate kitchen canisters, platters, plates, mugs, etc. Search for *mini decals*, which will bring up jewelry-sized selections.

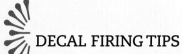

DECAL FIRING TIPS

- Cut closely around images with scissors or an X-ACTO blade.
- Place decals into lukewarm water for between 45 and 60 seconds.
- Carefully slide a decal from its backing sheet and onto your enamel.
- Smooth out any bubbles or wrinkles with your fingers and blot excess water with a soft cloth.
- Allow to dry for approximately three hours. If you're the impatient type (like me), slowly heat your piece in the nonvisible part of the flame to allow the water to evaporate and the enamel to gently warm. This process could take two to three minutes. If you rush this step, steam will build up under the decal and cause a blowout of your decal pattern; your enamel could experience thermal shock and crack. When the risk of steam and thermal shock has passed, bring the enamel piece into the hot spot of the flame.
- When possible, direct the flame away from the decal. For instance, in the case of a pendant, direct the flame to the back side. In the case of a bead, it's not possible to avoid the decal with the flame.
- When the metal appears to be hot enough that enamel will stick to the surface, gently lay the piece face down into 2020 Clear for Silver or 2030 Medium Fusing Clear enamel. You may also sift the enamel onto the piece to prevent the decal from shifting. One layer of enamel is all that is needed to protect the decal.
- Heat the surface until smooth. Do not heat longer than is necessary or the detail of the image will fade (in the case of a homemade decal) or become blurred (in the case of a commercial decal).

Making Enameled Head Pins, Twisty Tendrils, Pressed Flowers and Free Form Flowers

Making Enameled Head Pins

Head pins are a staple of jewelry making. With enamel, you can add little bits of color to your compositions by using enamel head pins. If you've shopped around for enamel head pins, you already know how costly they can be. If you flub up when using one, your costs have just doubled. So let's learn how to make them, shall we?

TORCH-FIRED TIP

The secret to creating a head pin is to get the tip of the wire in the hottest part of the flame and be aggressive about melting the end of the wire. Do not dillydally. The more oxidation that builds up on the wire, the harder it is to ball up the wire.

The opposite is true of melting the enamel on the metal head pin. You want to caress the glass with the flame. The glass naturally wants to round out. Don't blast it into shape, but coax it into shape.

 Place 6/20 enamel in a heat-resistant container. Cut 3" (8cm) wires from 20-gauge or 22-gauge wire. This is where a pair of cross-locking tweezers is appreciated because they stay closed when they're at rest; you do not need to squeeze them closed. Start the torch.

2 Dangle the loose end of the wire in the sweet spot of the flame. The end of the wire will melt and ball up. As the ball draws up, lower the wire further into the flame. You'll watch the melted ball get larger. It's as if you're chasing the melted ball up the wire.

3 Because it is imperative that the melted wire be very hot as it touches the enamel, you will need to hold the enamel container 4"–5" (10cm–13cm) below the end of the torch nozzle. Enamel will only stick to hot metal. If the enamel does not stick, reheat and try again.

4 With a chip or two of 6/20 enamel stuck to the end of the melted ball of wire, bring the enamel into the flame. When the enamel melts and rounds out, repeat the dipping and heating process until you're satisfied with the size of your glass head pin.

5 You're not quite finished with the head pin because you need to do flame annealing. Flame annealing is a process of controlled cooling that helps to prevent thermal shock and cracking. Hold the head pin in the nonvisible part of the flame and watch the glow fade from the glass. Set the head pin in your Bead Pulling Station. Now you're done.

Making Twisty Tendrils

Twisty tendrils are great for hiding the mechanics of your work, like the crimp bead and jump rings associated with a clasp. They are also useful for a bit of standout color in a composition. They don't require space on a cord or beading wire, which makes them flexible in more ways that one!

The procedure for creating twisty tendrils is the same as for creating enamel head pins except that you will add enamel to both ends of a 5"–6" (13cm–15cm) length of 22-gauge wire. After the enamel has cooled, simply wrap the wire around a pencil, mandrel or other cylindrical object.

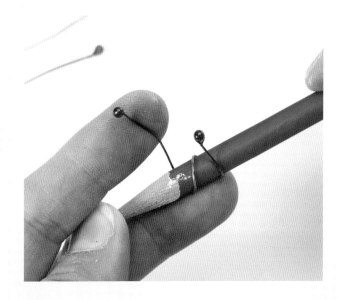

Enameling Flower Head Pins

Creating flowers with the Carlo Dona Bellflower Press is a breeze. If you like your flowers a little less uniform, however, you can pinch out your own with a pair of tweezers! Both processes take a little practice like most things that you want to do well.

The procedure is the same as for creating enamel head pins except we're going to need a crock pot to heat vermiculite for slower cooling and we're going to need a larger gather of glass on the end of the wire.

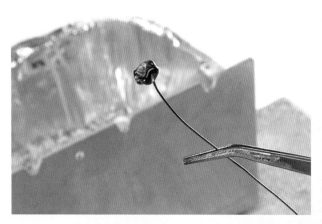

1 Create an enamel head pin with 6/20 enamel until you have a head that measures about 8mm. After you've gathered enough glass on the end of the wire, wait a second or two before turning the wire upright. The ball of glass atop the wire looks like a lollipop.

2 Slide the wire into the slot of the bellflower press, squeeze the jaws closed and quickly release. Wave the flower in and out of the flame. Be careful not to heat the glass so much that you lose the definition of the flower. Flame anneal (see *Making Enamel Head Pins*, step 5) and then place the flower head pin in a crock pot of heated vermiculite.

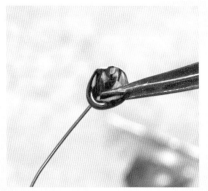

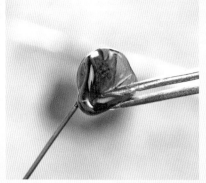

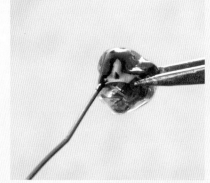

3 If you don't have a press, simply use a pair of tweezers or chain-nose pliers to pull and form petals.

A pressed glass flower in the
Forest Nymph composition.

Firing Transparent Enamels

Transparent enamels help us, as artists, to enhance the shape and form of the metal we're enameling. Oxides that develop when metal is heated require that we give special consideration to the enameling process when trying to preserve the true color and clarity of transparent enamel. This section deals primarily with firing transparent enamels.

Enameling Brass

Applying transparent enamel over brass enhances the qualities of both materials. The light golden color of the brass encourages the transparent enamel to look its best. But to get to that end point, there are a few rules we should follow. A little understanding of the dynamics at work will go a long way toward attaining our desired result.

When metal is in the presence of oxygen, oxides form on the surface. This oxidation process is accelerated when the metal is heated with a torch. This is what makes the metal look dark when it is heated. Opaque enamels simply cover the dark metal. However, this oxidized metal will severely affect the fired color of transparent enamel.

The goal is to reduce the oxides formed during the initial heating process and to trap those that do form in clear enamel (also known as flux). When firing brass, keep your piece just beyond the sweet spot of the flame—in the cooler part of the flame—and heat to a dull cherry red. Get the piece just hot enough so that enamel will stick to it. Quickly dredge the brass through flux. The flux will seal off the metal and prevent the formation of more copper oxides. But

Brass bead enameled in 2530 Water Blue.

more important, it traps the copper oxides that have already formed.

When we continue to gently heat the brass, the copper oxides will be absorbed by the flux. As this process is taking place, you'll see the brass transform before your eyes—from a dark piece of metal to a golden one. Now is the time to apply any number of beautifully colored transparent enamels. Allow the transparent enamel to accentuate the texture or details of your brass piece. Please remember that transparent enamels stay tacky a second or two longer than do opaque enamels, so allow extra time on the mandrel before removing at the Bead Pulling Station.

Enameling Copper

With opaque enamels, simply fire as usual. However, if you like the clear, juicy color of transparent enamels, you'll need to make some adjustments to the firing technique. As with brass, oxidation builds up on copper when heated. We can use the same principles used to fire brass, specifically, trapping copper oxides in clear enamel, burning them out through a gentle and slightly extended firing, and dredging the metal through colored transparent enamel. However, I've never been able to burn out enough of the copper oxides to produce a result that reflects the true color of the transparent enamel.

The solution would be a hybrid approach that embraces traditional enameling and the Immersion Process. Unfortunately this would involve cleaning a little metal and spraying an adhesive, but we would retain some of advantages of the Immersion Process.

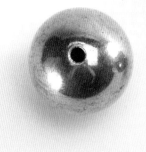

Overfired copper bead with 2030 Medium Fusing Clear.

Lay paper on your work table. Clean the copper bead with a copper cleaner like, Penny Brite. Brush it with an adhesive, like Klyr-Fire (diluted 1:1 with water) .

Place the bead on a mandrel. While holding the bead over the paper, sift enamel onto the bead. Allow the adhesive to dry. Heat the bead in the flame and add subsequent layers of enamel using the Immersion Process.

Enameling Iron

Fire iron beads as you would brass. My son calls this "the slow-roasting" method. With clear enamel, you can sometimes get an iridized look to the metal. It's very exciting!

Iron bead enameled in 2030 Medium Fusing Clear.

Annealing

Heating copper to a dark red is called annealing. Annealing makes metal pliable and easier to shape, in a dapping block or by formfolding. As you work metal using processes such as hammering, bending, shaping and folding, the metal becomes work hardened and resistant to shaping. Annealing will return the metal to a pliable condition.

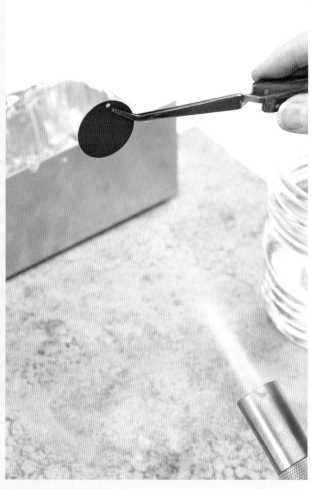

1 Hold the metal with cross-locking tweezers or old pliers. Don't use your good ones! Place the metal in the flame until it turns a dark red.

2 Remove the metal from the flame and quench it in water. Dry it with a cloth or paper towel. Don't allow wet metal to come into contact with steel tools!

Soldering

Warning! This is a simplified discussion of soldering. Entire books are written on the subject because there are many ways to solder, many types of solder and many temperatures at which they flow. Hopefully this is enough information to get you started. The two most common types of soldering are hard soldering for fine jewelry and soft soldering for plumbing. Thanks to Stephanie Lee, who pioneered the use of soft solder as a 3-D element in her gorgeous jewelry, soft solder has taken a walk from the hardware store to the jewelry store next door.

Some commonalities exist, of course. Both types require the metal be clean and the area around the joint to be painted with flux to help the solder flow. But that's pretty much where the similarities end.

Directions for Hard Soldering

1 Clean the area to be soldered with a scrubby pad and a liquid soap, like Dawn, that has no moisturizers. Rinse and dry.

2 Scrub the end of a sheet of solder. (Extra-easy solder melts at the lowest temperature).

3 To cut a small chip, or pallion, of solder, cut into the solder sheet as if you were creating fringe. The cuts should be very close together (about 2mm). Come in from the side and cut a pallion off the sheet. Ideally you want square pallions.

4 Paint paste flux on the area to be soldered. You may also use a drop of liquid flux if you prefer.

5 With tweezers, place the pallion at the seam. Since solder will flow toward the heat (flame), take this into consideration. You may want to place the solder under the seam so it can flow upward toward the flame. Hold the torch (a Butane torch is fine for this) in your nondominant hand so that your dominant hand is free to nudge the solder into place, if necessary.

6 Begin heating the metal by slowly circling the torch around the metal. The entire piece of metal needs to be heated, even though you may be soldering only one small area. Gentle heating also allows the water in the flux to evaporate. If you heat too quickly, the solder can jump off the seam. You will need the fine motor control of your dominant hand and a soldering pick or tweezers to nudge the solder back into place.

7 As the water evaporates, the flux becomes gray and dusty looking. At this point, you can direct the flame at the solder joint. When the flux starts to liquefy, direct the flame to the joint. Liquefaction is an indication that the solder will begin to flow soon. When the solder melts, continue heating until it flows. Immediately remove the torch from the seam. Allow the solder to cool for a few seconds before moving it with tweezers.

8 Pick up the piece with the copper tongs. Give it a quick dip in water. Place the piece in a pickle (a mild acid solution; see next page) to remove oxidation and excess flux. Remove the metal from the pickle with copper tongs, rinse it in water, and dip it into in a neutralizing bath of 1 teaspoon of baking soda to ½ cup of water.

9 Remove the piece from the bath, rinse and dry. Examine the seam. Give it a tug. If there is a raised bump at the solder seam, either the solder didn't flow or too much solder was used. If the solder didn't flow, reheat until it flows. Otherwise file the raised area with a needle file.

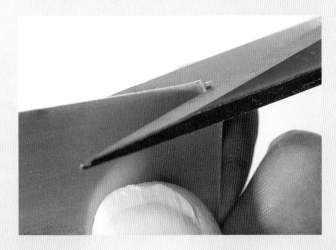

Pickle

Pickle … What is it? How do you use it? Why do we need it?

Pickle, a mild acid solution of sodium bisulfate and water, cleans metal of oxidation. Oxidation is a naturally occurring process that is accelerated by the heating of metal. Oxidation acts as a barrier in certain processes, such as applying enamel to *cold* metal and soldering metal. In enameling, it prevents the adhesion of the enamel. In soldering, it prevents the solder from flowing.

Pickle can be purchased from jewelry supply houses. I use pH Down, which is a chemical used in swimming pools. Pickle is usually kept warm because warm pickle decreases cleaning time. Many jewelry artists keep pickle in a Crock-Pot. For a 1½ quart crock pot, add ¾ cup of dry pickle to 4 cups of water. Always add acid to water, never the other way around.

Always use copper, bamboo or plastic tongs with pickle. If the pickle is used for both sterling and copper jewelry, retrieving jewelry from the pot with steel tweezers will result in the sterling being plated with copper, potentially ruining your tweezers.

To use the pickle, rinse the metal and add it to the pickle with tongs. When the metal is clean (this usually only takes a few minutes), remove it from the pickle with tongs. Rinse the piece in clear water and neutralize it in a solution of ½ teaspoon of baking soda to 1 cup of water. Rinse once again in clear water and dry.

Alternate Ways to Remove a Pendant

If you were to remove the pendant by pulling it off the mandrel, as you would a bead, the excess enamel on the mandrel would fill the pendant hole. Here are two options for avoiding the pitfalls of removing work when excess enamel has built up on the mandrel.

Excess enamel builds up on the mandrel during the firing process, particularly when firing flat pieces that need to be located further from the end of the mandrel during firing. One way to remove a pendant or flat piece is to allow the enamel to cool on the mandrel. Crack the enamel off with old pliers or wire cutters. Heat the connection point between the piece and the mandrel. Pull the piece off the mandrel.

The other option is to push the pendant off the back end of the mandrel. To do this, simply place the mandrel in the V notch of the Bead Pulling Station. With the pendant on the outside of the BPS, push your hand away from your body. The pendant will move down the mandrel toward your hand and be freed from the excess enamel. With pliers, grab the hot end of the mandrel and shake the pendant off the cool end of the mandrel.

TORCH-FIRED TROUBLESHOOTING

ENAMEL CHIPS OFF
- Metal was not heated sufficiently for the enamel to adhere.
- Get the metal hotter, especially for the first layer.
- Bead was dropped onto a hard surface.
- Uneven heating of large objects.
- The metal is incompatible with the enamel.

BEAD STICKS TO MANDREL
- On the first pull, try to get the bead so that it is dangling on the end of the mandrel. If the bead sticks, simply heat the hole at the end of the bead closest to your hand.
- If the mandrel is protruding from the other end of the bead, reheat the entire bead to release from the mandrel. No amount of pulling will release the bead. Heat is the answer.

ENAMEL PULLS FROM ONE END OF BEAD AND HAS SHARP EDGES
- The enamel at the join of the mandrel has cooled faster than the enamel on the bead. Do not continue to pull when you feel resistance. Simply heat the end of the bead closest to your hand.

UNEVEN ENAMEL APPLICATION
- Uneven heating of bead—reheat the bead evenly. Rapidly rotating the mandrel in the flame helps.

BUBBLES IN THE ENAMEL
- The enamel has been overfired or the metal is incompatible with the enamel.

ENAMEL IS ROUGH
- Enamel was not fired hot enough—reheat and rotate evenly until the bead glows orange.

BEAD IS DARK
- When overfired, enamel pulls away from the edges, leaving a dark lining around each opening in a bead or around the edges of sheet metal. Refire and add more enamel.
- Beads can also become muddy when fired too close to the end of the torch nozzle. Remember to work in the sweet spot of the flame.

METAL IS WARPED
- Counter-enamel is enamel that is on the back side of your flat piece. Having enamel on both sides of the metal creates equal compression, which prevents warping of the metal. It is particularly important to flat pieces, which are prone to warping. Warping will cause the enamel to chip off the surface of the metal. You can buy a product called counter-enamel from your enamel supplier. It is usually dark enamel applied to the back of your piece. However, counter-enamel is a natural by-product of working in the studio. When you clean your work area at the end of the day, all of the colors you brush off into a dish will create a dark enamel that you can use for the back of your piece. However, since we're applying enamel to the front and back of our flat pieces at the same time when we dredge the piece through enamel, counter-enameling is not a separate step for us!

For bonus projects and more visit: CreateMixedMedia.com/Mastering-Torch-Fire.

35

Color Development: Layering Enamels

The best way to create new enamel colors is by layering transparent enamels over opaque enamels. However, you want to make sure that the metal color doesn't interfere with the final result. Here are a few guidelines for enameling different metals.

IRON

I recommend using two layers of opaque enamel as a base to cover the dark color of the metal (or use the same firing method as brass). As you can see from the color recipe for Tangerine Tango, a recent Pantone Color of the Year, base layers of enamel need not be limited to white.

COPPER

When we heat the bead, we create copper oxides, which are dark. This will severely affect the color of transparent enamel. We have two options. We can apply an opaque enamel as a base or we can using the hybrid firing method recommended on page 30. However, it is possible to create beautiful transparent color over copper, though it generally will be a darker version of the color selected.

BRASS

You can skip using opaque enamels as a base. Transparent enamels will look luscious and true to color over brass. Transparent enamels will pool in the depressions of the designs of brass stampings, creating a gorgeous detailing of the design.

The following are some of our favorite color recipes at Painting with Fire Studio. What you'll discover is that the final color created is a logical progression of the layering of enamels, much like layering crayons in a coloring book. Have fun in the studio and create some of your own signature colors! Here are the recipes for plenty of new colors. (The recipe for each color is listed top to bottom, first layer to last, as shown in the photos.)

Red Violet
Bead 1: 2 layers 1055 White, 1 layer 2836 Raspberry

Bead 2: 2 layers 1055 White, 1 layer 2650 Heron

Bead 3: 2 layers 1055 White, 1 layer 2650 Heron, 1 layer 2836 Raspberry

Salmon
Bead 1: 1 layer 1055 White, 1 layer 1715 Clover

Bead 2: 2 layers 1055 White, 1 layer 2850 Woodrow Red

Bead 3: 1 layer 1055 White, 1 layer 1715 Clover, 1 layer 2850 Woodrow Red

Mustard
Bead 1: 2 layers 1055 White, 1 layer 2215 Egg Yellow

Bead 2: 2 layers 1055 White, 1 layer 2240 Olive

Bead 3: 2 layers 1055 White, 1 layer 2240 Olive, 1 layer 2215 Egg Yellow

Caribbean
Bead 1: 2 layers 1055 White, 1 layer
2230 Lime Yellow

Bead 2: 2 layers 1055 White, 1 layer
2520 Aqua Blue

Bead 3: 2 layers 1055 White, 1 layer 2520
Aqua Blue, 1 layer 2230 Lime Yellow

Tangerine Tango
Bead 1: 3 layers 1830 Marigold

Bead 2: 2 layers 1055 White, 1 layer
2850 Sunset

Bead 3: 2 layers 1055 White, 1 layers 1830
Marigold, 1 layer 2850 Sunset Orange

**An example of the darkening effects
of layering complementary colors**
Bead 1: 3 layers 1870 Orient Red

Bead 2: 2 layers 1870 Orient Red, 1 layer
2305 Nile Green

An example of saturation
Bead 1: 2 layers 1055 White, 1 layer
2836 Raspberry

Bead 2: 2 layers 1055 White, 2 layers
2836 Raspberry

Bead 3: 2 layers 1055 White, 3 layers
2836 Raspberry

The Projects

The more I work in the studio, the more in awe I am of the possibilities of torch-fired enameling. The fact that I'm using an inexpensive torch to create artwork delights me still! The Immersion Process has definitely brought enameling to the kitchen table. The tools are inexpensive, the small space requirements are easy to meet, the technique is easy to learn and the tedious steps—the things that discourage people from ever getting started—have been eliminated. But we must acknowledge that to learn something new requires at least a little patience and persistence.

What follows is a not-uncommon experience in the Painting with Fire teaching studio. I start all workshops by giving instruction, of course. Everyone starts by enameling about six to eight small iron beads. After a few minutes I start peeking into bread pans. I can evaluate how the students are doing pretty quickly by how their beads look. I knelt down to talk to one student and she expressed how discouraged she was. I pointed to one of her beads and said, "Okay, this was probably your first bead. You can see that the end didn't get covered. These two are better, but these two are perfect!" She looked up at me as I continued, "Remember, you've just been working at this for ten minutes!" and we both laughed.

I love the following quote by radio personality Ira Glass. It reflects not only my personal experience as an artist working in her studio, but also my experience with students:

"All of us who do creative work, we get into it because we have good taste. But there is this gap. For the first couple years you make stuff, it's just not that good. It's trying to be good, it has potential, but it's not. But your taste, the thing that got you into the game, is still killer. And your taste is why your work disappoints you. A lot of people never get past this phase, they quit … We all go through this. And, if you are just starting out or you are still in this phase, you gotta know it's normal and the most important thing you can do is do a lot of work …"

My goal in writing this book is to share with you the discoveries I've made by doing a lot of work over the past three years. But remember, "You gotta do the work, close the gap, and don't quit!" Now get out there and do it!

MATERIALS

Enameling Tools
basic torch-fired enameling kit

Enamel
1830 Marigold
1860 Flame Orange
1995 Black

Metalworking Tools
bead reamer or needle file
chain-nose pliers, 2
chasing hammer
hole punch
metal shears
rubber bench block
wire cutters

Other Materials and Findings
beads, Arabesque filigree, 20mm
beads, Pierced Temple filigree, 14mm
bead caps
clasp, ball-and-hitch
copper sheet, 24-gauge
cotton-filled cording, 12" (30cm)
head pins
jump rings, 9mm
sifter

TECHNIQUES

Firing Flat Objects
Firing Delicate Shapes
Pendant Removal Options

Sunflower of Provence

A trip to Aix-en-Provence is definitely on my bucket list, but until I get there, this necklace is a worthy tribute! We'll cover the end of the fabric cording with delicate enameled bead caps and draw a stylized sunflower using a sgraffito technique.

1 Cut a 1⅞" × 1½" (48mm × 38mm) oval from a sheet of copper.

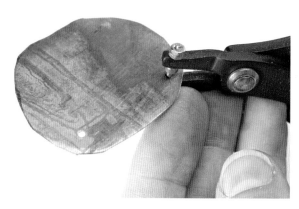

2 Punch 2 holes near the top of the pendant.

3 Hammer the piece so it is slightly domed.

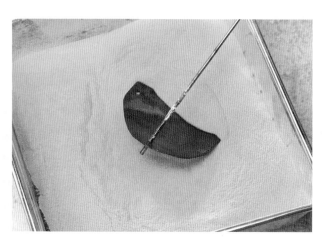

4 Enamel the pendant in 3 layers of Marigold enamel.

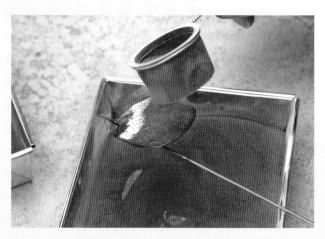

5 While the piece is still hot, sift 1 layer of Flame Orange onto the surface. Fuse the enamel.

6 Dredge the entire pendant in Black enamel. Do not fire. Cool the pendant slightly to allow the pendant to become stuck to the mandrel.

7 The mandrel now becomes a handle by which you can turn the pendant. (You can also hold the pendant steady with pliers, if you prefer.) Using a bead reamer or needle file, scratch a sunflower design through the black enamel.

8 Place the pendant back into the flame and heat it until glowing. Remove it from the mandrel.

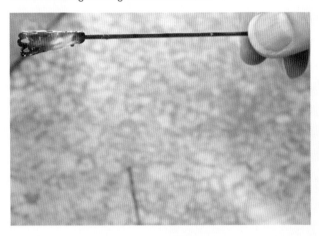

9 Enamel the bead caps in Flame Orange.

10 Enamel 6 large beads and 2 small beads in Flame Orange.

11 Thread the enamel beads onto head pins and create a simple loop at each end of the bead.

 JEWELRY-MAKING TIP

Use stiff wire for simple loops and dead soft wire for wire wrapped loops.

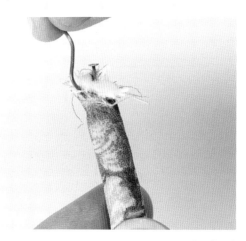

12 Link 3 large beads and 1 small bead together with jump rings. Repeat with the remaining beads so you have 2 chains each comprised of 4 beads.

13 Pierce the end of the cotton-filled cording with a head pin. Fold the ends of the wire up and wrap the beaded end of the head pin around the wire.

14 Thread the extending wire through the enameled bead cap. Place a faceted copper bead on the wire and create a simple loop. Repeat steps 13–14 with the other side of the necklace.

15 Use a jump ring to join the fabric and beaded parts of the necklace. On 1 side, connect the fabric and beads with a jump ring, and on the other side, attach the ball-and-hitch clasp.

16 Attach the focal piece to the beaded strands with jump rings.

MATERIALS

Enameling Tools
basic torch-fired enameling kit

Enamels
1175 Mocha

1415 Sea Foam

1860 Flame Orange

2030 Medium Fusing Clear

Metalworking Tools
chain-nose pliers, 2

chasing hammer

hole punch

metal file

metal shears

needle file

rubber bench block

Other Materials and Findings
brass flowers, large, with at least 3 petals per flower, 2

chain, copper-plated, 2mm × 3mm

container with water

earwires, copper-plated, lever-back, 2

jump rings, copper-plated, 5mm

decal sheet, harlequin pattern

marker

TECHNIQUES
Firing Flat Objects

Firing Decals

Pendant Removal Options

I'm Distressed Earrings

I must admit I have trouble getting dangles to face the right way when I hang them from a chain! Here's a tip: If your dangle faces the wrong way, do not take the jump ring out of the link! Simply open the jump ring and reverse the orientation of the dangle. In this project we'll layer opaque enamels and scratch unfired enamel from the edges to create a distressed look. Finally, we'll apply and fire ceramic decals.

1 Cut the flower petals from the brass flower. Separate 3 pairs of petals to make 2 earrings. Smooth rough edges with a needle file.

2 Place the petals on top of the rubber bench block and begin hammering the petals. You'll notice that the petals will cup upward.

 TORCH-FIRED TIP

Because the metal is so thin, you won't need to anneal it before forging the petals. The hammer marks also will add to the charm.

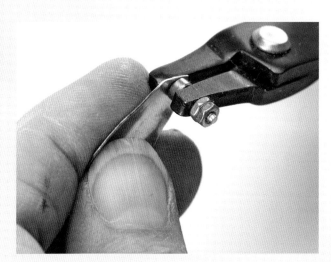

3 Punch a hole at the top of each petal. File edges. Remember to keep the hole close enough to the edge to accommodate the size of your jump ring.

4 Enamel 1 of the petals with 2 layers of Mocha enamel.

5 Fire and dredge the petal through Sea Foam enamel. Before fusing the enamel, scrape Sea Foam enamel from the edges with a mandrel or other metal object such as a file, revealing the Mocha.

6 Fire the petal from underneath to fuse the enamel because the Sea Foam enamel is loosely sitting on top of the petal. Repeat steps 5 and 6 one more time.

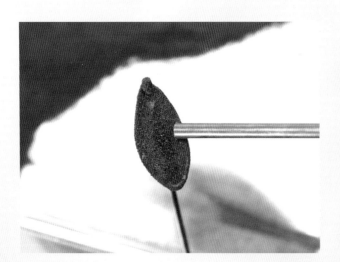

7 Repeat the process until you have 4 petals with Mocha and Sea Foam enamels and 2 petals with Mocha and Flame Orange enamels.

8 Lay a petal on the decal paper and trace the outline of the petal onto the paper. Cut out the petal shape. Repeat for the remaining petals.

9 Soak the decals in water for 45–60 seconds, then slide the decals off the backing paper and apply them to the petals. Smooth the decals with your fingers, and blot any excess moisture from the decals.

10 Fire a petal. Then add a layer of Medium Fusing Clear, and fire again. Refire the petal by heating it very slowly in the nonvisible part of the flame. This will evaporate the moisture of the decal and prevent steam from building up beneath the decal.

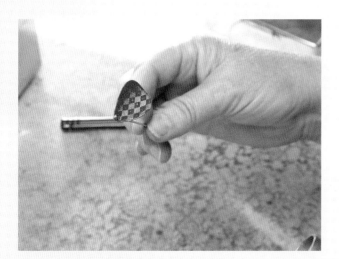

11 Add another layer of clear enamel, and fire again. Repeat steps 10 and 11 for the remaining petals.

12 Cut two 1¼" (3cm) pieces of chain. Use jump rings to attach the dangles to the chain. Attach the chain to earwires.

MATERIALS

Enameling Tools
basic torch-fired enameling kit

Enamels
2030 Medium Fusing Clear

1055 White

1870 Orient Red

Metalworking Tools
bench block, steel

chain-nose pliers

chasing hammer

Crock-Pot

vermiculite

wire cutters sufficient to cut
16-gauge wire

Other Materials and Findings
clasp, vintage hook from hook
and eye closure

container of water and baking
soda solution

cylinder shape (such as a dowel
or dapping punch), optional

darkening agent, such as Black
Max

leather cording, 2mm

paintbrush

sanding pad

wire, rebar, 16-gauge

wire, sterling silver, 16-gauge

Rings of Fire

How many enamel links can you get from 335' (8.5m) of 16-gauge wire? Let's just agree that the answer is a lot. But what if I told you that the wire would cost less than $6 and you wouldn't even need to clean it before enameling? Would you be even more interested? If so, this is the project for you … and me! When you go to the hardware store, rather than ask for black wire or annealed steel wire, simply say, "I'd like to buy some rebar wire." They will direct you to the masonry area, where you'll find some black wire that will immediately make your hands dirty. But soon, with very little effort, it will become bright red enamel!

TORCH-FIRED TIP

Put vermiculite in the Crock-Pot and turn it to high. You will need it to be heated for step 9.

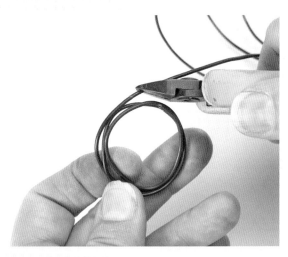

1 Create 8 rebar wire links, 4 that are approximately 1¼" (3cm) in diameter and 4 that are approximately ¾" (12cm) in diameter. Each link should consist of approximately 3 wraps of the wire each.

2 Try to keep the wire loops as close together as possible to avoid large gaps that enamel would have to span.

METALWORKING TIPS

• You can create all the links free form or you can wrap the wire around a dowel or other cylinder. I use the end of my cool torch as a form.

• Bevel-cut each end of the wire and tuck the wire ends into the loops to prevent sharp edges.

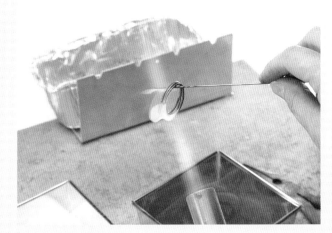

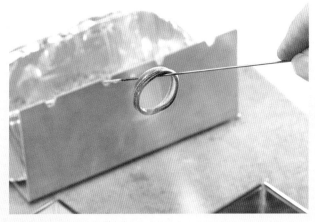

3 Dangle a wire link from a mandrel. Heat the link in the flame until it glows orange.

4 Dredge the link in Medium Fusing Clear. Fire the link.

5 Apply 2 to 3 layers of White, reheating between layers.

6 Apply 2 to 3 layers of Orient Red, reheating between layers.

7 When you're satisfied with the color, remove the link from the mandrel by tilting the tip of the mandrel toward your work surface and rotating the mandrel. Focus the flame on the connection point between the link and the mandrel, while avoiding heating other areas of the link.

8 When the link is close to coming off the mandrel, use pliers to grab the cool part and remove it from the mandrel. Use the flame to heal over the connection point of the enamel.

9 Place the link in heated vermiculite or Oil-Dri. Repeat steps 1–9 with the remaining 7 rebar links. When all links are enameled, turn off the Crock-Pot and allow the links to cool gradually. Do not rush the cooling process.

10 Create 20 jumps rings, each 1" (25mm) in diameter, from 14-gauge sterling silver or nickel silver wire.

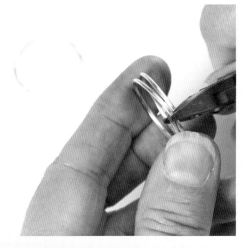

11 Make another 20 sterling silver or nickel silver jump rings; this time each should be approximately ¾" (19mm) in diameter.

12 Lay a jump ring on top of a bench block. Flatten it on 1 side with the flat side of your hammer. Flip the link over and repeat on the other side.

METALWORKING TIP

It is natural for the 2 ends of the link to separate during the forging process. When this happens, bring the ends together and continue forging.

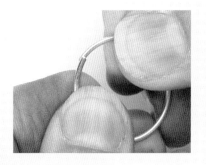

13 Forge both sides of the jump ring again, but this time use a the ball-peen side of your hammer to create light-reflecting facets in the wire.

14 Oxidize the wire with Black Max or other darkening agent.

15 Place it in a solution of baking soda to neutralize and then rinse it in water.

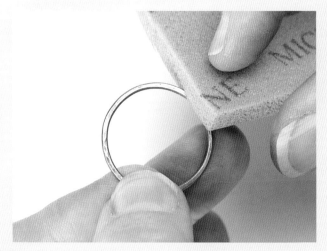

16 Buff with a scrubbing or sanding pad to bring out the highlights of the texturing. Repeat steps 12–16 with the remaining 39 jump rings.

17 Start laying out your necklace design, spacing out enameled links. Connect the links with a combination of large and small jump rings. Add extra jump rings to add interest to the design.

18 When you get to the focal area, add extra jump rings in order to allow the enameled links to create an asymmetrical design.

JEWELRY-DESIGN TIP

A jewelry bust can be an aid when designing your necklace.
T-pins are handy for securing the design while you work.

19 Cut 30" (76cm) of leather cording. Double it by folding it in half. Pass 1 end through the last link on 1 side of the necklace. Tie a Lark's head knot and an overhand knot.

20 Pass the other end of the leather cording through the 2 openings of the hook assembly. Tie an overhand knot. Clip off any excess cord.

MATERIALS

Enameling Tools
basic torch-fired enameling kit

80 Mesh Enamel
1410 Robin's Egg

2030 Medium Fusing Clear

2305 Nile Green

BC303L Clear Liquid Enamel, custom tinted

6/20 Enamel
2410 Copper

2305 Nile Green

Metalworking Tools
bench block, steel

chain-nose pliers

chasing hammer

cross-locking pliers

dapping block and punch

hole punch

needle file

round-nose pliers

wire cutters

Other Materials and Findings
beads, briolette and faceted

clasp, sterling ball and chain

copper disc with holes on opposite sides, 1¼" (3cm)

head pin, silver , 20-gauge

jump rings, 4mm and 8mm

mica flakes

paintbrush

sea glass

tweezers

wire, sterling silver, 24-gauge

TECHNIQUES

6/20 Enamel

Liquid Enamel

Firing Flat Objects

Firing Transparent Enamel

Annealing

Pendant Removal Options

Tide Pool

With **Tide Pool**, we're capitalizing on the glittery qualities of mica and adding liquid enamel that has been tinted a beautiful Bermuda Green. The liquid enamel adds a juicy depth to the tide pool and provides a liquid into which the mica flakes can sink. You'll also find a lesson on how to wire-wrap sea glass!

1 Anneal, quench and dry the disc. Slightly dome the copper disc in a dapping block. Fire the disc with 2 to 3 layers of Robin's Egg. Allow the disc to cool and stick to the mandrel so you can use the mandrel as a handle.

Paint a puddle of tinted liquid enamel in the center of the disc.

2 Lay mica flakes on the liquid enamel and press them in.

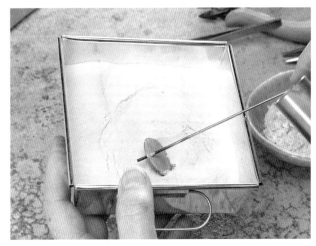

3 Bring the pendant back into the flame. Fire gently from underneath to allow the enamel to warm gradually and the moisture from the liquid enamel to evaporate.

4 Dredge the disc through Medium Fusing Clear enamel. Heat to glowing. Repeat.

 TORCH-FIRED TIP

Mica is archival, mined from the earth, heat-resistant and looks much like silver when applied to enamel. It comes in various forms including powder, chips, flakes and sheets. Please don't confuse glitter for mica or you'll have an ooey, gooey mess upon heating!

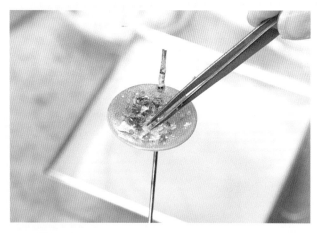

5 While heating the pendant from underneath, apply 6/20 enamels in Copper and Nile Green.

6 When the enamel begins to slump and shows signs of fusing, bring the flame topside to smooth the rough edges of the 6/20 enamels. Remove the pendant from the mandrel and allow it to cool.

7 For the enameled wire dangle, cut approximately 3" (8cm) of 16-gauge sterling. Flatten 1 end with a hammer. You are also stretching the wire as you flatten it. Instead of hammering up and down, hammer downward and toward the end of the wire to thin it.

8 Use a hand punch to create the hole.

 METALWORKING TIP

Be sure to flip over the wire so it doesn't curve in any one direction. File the ends smooth with a metal file.

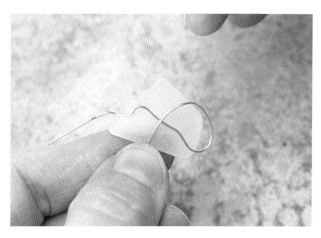

9 Hold the flattened end of the wire with cross-locking tweezers or pliers. Heat the wire and dip it into Nile Green enamel. Repeat 1 or more times and then cool.

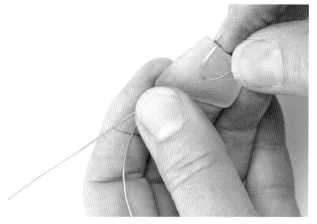

10 To gauge how much wire you'll need to wire-wrap the sea glass dangle, hold one end of 22-gauge wire and loosely create a cross on one side of the sea glass. Measure and then triple this measurement. Cut the wire.

11 To wrap the sea glass, grab the center of the wire with round-nose pliers. Press both sides down. Lay the loop on top of a bench block and hammer to flatten the bend.

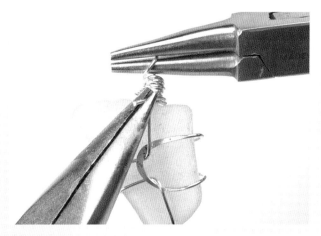

12 Lay the bend on the center of the front of the sea glass and fold the wires to the back of the sea glass.

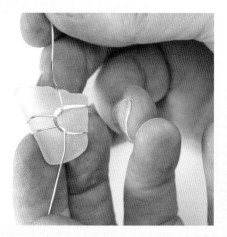

13 Bring both ends of the wire to the front. Bring the ends up through the loop. Hold the top wire in your dominant hand. Wrap the other wire around the back of the glass

14 Cross the two wires at the top edge of the sea glass and create a wire-wrapped loop.

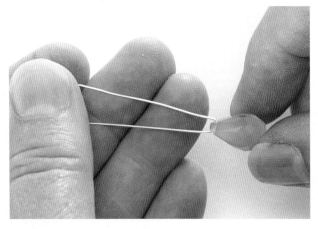

15 To create the briolette dangle, cut 4" (10cm) of 24-gauge wire. Pass the wire through the hole in the briolette until the bead reaches the midpoint of the wire. Bend both ends of the wire upward.

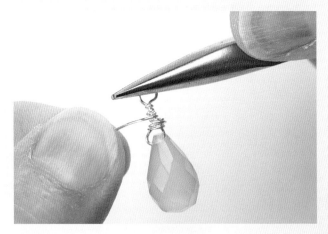

16 Grab 1 end of the wire with pliers and wrap it with the other wire. Create a wrapped loop at the top.

17 Put all the dangles on an 8mm jump ring and add the jump ring to the pendant.

18 Put the pendant on the chain.

EFFERVESCENT REFLECTION
For a downloadable PDF with instructions for making **Effervescent Reflection**, visit createmixedmedia.com/mastering-torch-fire.

Spectator

MATERIALS

Enameling Tools
basic torch-fired enameling kit

Enamels & Enamel Supplements
1055 White Enamel

P-1 Black Overglaze Enamel

Metalworking Tools
brass mallet

chain-nose pliers, 2

dapping block and punch

die-cut machine or decorative
punch

Other Materials and Findings
adhesive-backed vinyl

chain, various styles

clasp, ball-and-hitch

copper discs, 1 " with ³⁄₃₂" holes,
annealed, 5

jump rings, 8mm and 14mm

marker

mineral spirits

paintbrush

paper

TECHNIQUES
Overglaze Enamel

Annealing

Pendant Removal Options

Rows and rows of chain define boho chic jewelry. How about a little classic black-and-white thrown in? This project features a low-tech but nifty way to help you decide where to place the pendants on five rows of chain. All you need is a paper at least 26" (66cm) long. A newspaper would do it (if you can still find one!). Also, did you know that you can buy discs with the holes already drilled in them? They make life a lot easier.

1 Anneal all the discs and dap them for a gentle dome.

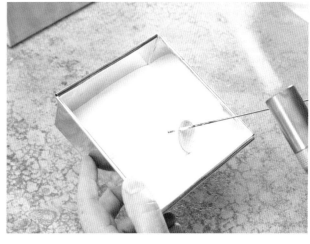

2 Enamel a disc in White enamel. Allow the disc to cool.

TORCH-FIRED TIP

If you don't have a dapping block to dome the discs, an inexpensive rubber bench block and ball-peen hammer is a great substitute. Lay the annealed disc on the rubber bench block and start hammering in the center of the disc. The disc will cup upward, toward the hammer.

3 Run adhesive-backed vinyl through a die-cut machine or use a decorative punch to create a stencil. Remove the paper backing and press the vinyl onto an enameled disc. Make sure to seal the edges of the design well.

4 Load a soft bristle brush with P-1 Black Overglaze enamel and paint over the stencil.

5 Remove the stencil and clean up any edges. Refire to fuse the P-1 Black Overglaze Enamel to the base enamel.

6 Repeat steps 2–5 for all discs.

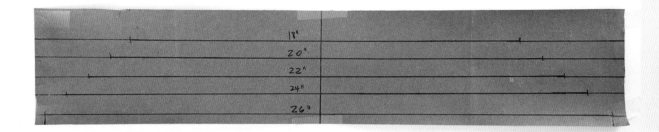

7 Directions for assembling your necklace are as follows: The longest chain in this necklace, measuring from the clasp, including the disc, is 26" (66cm). Each row diminishes in size by 2" (5cm). The easiest way to keep this straight and orderly is to create a paper template. Either tape pieces of paper together to create a sheet 26" (66cm) long or open a newspaper to get that length. Draw 5 parallel lines spaced 1" (24mm) apart. Fold the paper in half and draw a line on the centerfold. On row #1, the ends of the necklace will be meet the edges of the paper. For row #2, come in from the edges of the paper by 1" (25mm) on each side. For row #3, come in from the edges of the paper by 2" (5cm) on each side. Continue until all lines are measured and marked.

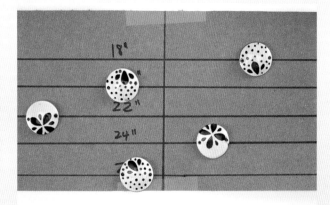

8 Cluster the discs toward the center of the necklace so a shirt collar or your hair won't obscure them. Place a disc on each row in a pattern that suits you.

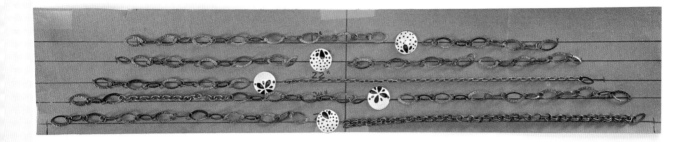

9 The remainder of the line that is visible indicates the length of chain you need for that row. To create more interest in your design, choose several different chain styles for your composition.

10 Connect chains and pendants with 8mm jump rings.

11 After all of the rows are finished, use a 14mm jump ring to gather the chains on 1 side of the necklace. Repeat for the other side.

12 Add a clasp to the necklace with jump rings.

MATERIALS

Enameling Tools

basic torch-fired enameling kit

80 Mesh Enamel

2030 Medium Fusing Clear

2335 Peacock Green

2520 Aqua Blue

2839 Red

2880 Woodrow Red

Metalworking Tools

chain-nose pliers

chasing hammer

flat-jaw welding pliers

file, diamond or needle

hole punch

metal shears

Other Materials and Findings

brass flowers, stamped, 22mm

brass sheet, textured

closure template

copper disc, 6mm

copper eyelets, 1.8mm

Crafted Findings Riveting Tool
marker

mini screws and nuts, optional

purse strap

tape

wire, copper, 14-gauge

TECHNIQUES

Firing Transparent Enamel

Enameling Brass

Up-Cycled Daisy Bracelet

Did you know that up-cycling is the process of converting so-called useless things into things of greater value? A purse strap and some cool hardware—primarily available to manufacturers— are given new life in this bracelet project. As a bonus, no special leatherworking tools are required!

1 Select a section of a purse strap or skinny belt for your
bracelet. Decide how you will open and close the bracelet. It
may be possible to use some of the hardware that is already in
place as part of your clasp or closure.

2 If you're unable to use the purse hardware and will be adding your own, mark its placement on the leather. Use metal
shears to cut the strap to your desired length.

3 Lay the flowers on the strap and mark their placement with
a marker.

4 Punch holes in the strap with a hole punch.

5 Enamel the brass flowers following the directions in the
Techniques section for enameling brass Enamel 1 each in
Peacock Green, Aqua and Woodrow Red.

JEWELRY-MAKING TIP

This is what your riveted daisy will look like from the front and back of the strap.

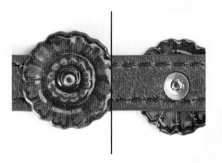

6 Place 1 of the enamel flowers on an eyelet and thread the eyelet through a hole in the leather bracelet. Use the end of your round-nose pliers to push the eyelet through the leather, if necessary. Place a 6mm copper disc on the eyelet. This disc will prevent the eyelet from pulling through the leather. Use the Crafted Findings Riveting Tool to compress the eyelet and secure the enamel flower to the leather.

METALWORKING TIP

You can use mini screws and nuts instead of the eyelets and riveting tool, if you'd like.

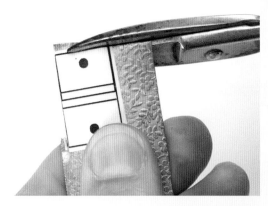

7 To make your own closure, tape the template (see Figure 4) to a brass texture plate. Use metal shears to cut out the shape.

8 Punch a ³⁄₃₂" (2mm) hole as indicated on the template. Leave the paper on or remove the paper for the next step. If you remove the paper, mark your fold line with a fine line marker.

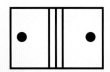

Figure 4: template

JEWELRY-DESIGN TIP

Many times I work out the kinks in a design by creating a paper template. However, because of the differences in the thickness of the two materials, leave a little extra allowance for the folds in the metal. You may enlarge or reduce this template at a copy machine to create one that will fit your project.

9 Hold the metal with the edge of the flat-jaw welding pliers aligning with the fold line. Hammer to crease the metal. Repeat for the other fold.

10 Cut a 1½" (4cm) piece of 14-gauge copper wire for the loop part of the closure. Shape the wire with chain-nose pliers. The width of the loop should be the approximate width of the brass tab. You are essentially creating a D-ring out of copper wire.

11 Slide the D-ring into the brass tab.

12 Position the tab on the leather strap and mark the hole placement. Remove the brass tab. Punch a hole in the leather.

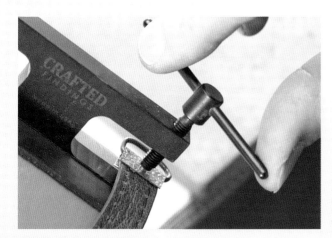

JEWELRY-DESIGN TIPS

In this particular bracelet we were able to use some of the purse manufacturer's hardware as part of the closure. However, if this is not an option for your project, create a second tab (steps 7–9) and another D-ring brass tab for the other end of the bracelet. Slide a large lobster clasp onto the D-ring. Follow steps 12–13 to finish the project.

13 Attach the tab to the leather with a copper eyelet. Use a Crafted Findings Riveting Tool or fasten the tab with micro nut and bolt.

Starlight

MATERIALS

Enameling Tools
basic torch-fired enameling kit

80 Mesh Enamel
2030 Medium Fusing Clear

2530 Water Blue

1410 Robin's Egg

1425 Sapphire

1430 Spruce

6/20 Enamel
1239 Mellow Yellow

2305 Nile Green

2410 Copper

1410 Robin's Egg

Metalworking Tools
bench block, rubber

chain-nose pliers

chasing hammer

hole punch

needle file

round-nose pliers

wire cutters

Other Materials and Findings
brass angel wing

chain, 2 styles

clasp, ball-and-hitch

copper discs, 10mm, 8

copper sheet, 24-gauge

jump rings, 8mm

TECHNIQUES
6/20 Enamel

Firing Flat Objects

Firing Transparent Enamel

Firing Brass

Pendant Removal Options

As we continue to explore the magic of 6/20 enamel, this time we will use a veritable rainbow of colors. This bracelet has a strong focal element, with a charm bracelet-style chain. A brass angel wing adds even more whimsy. Enameled brass sounds more like glass than any other enameled metal—you'll hear a little tinkle when the angel wing kisses the other enamel pieces on your bracelet!

1 Cut an organic shape, approximately 2½" × 1½" (6cm × 4cm), from 24-gauge copper.

2 Lay the copper on a rubber bench block and forge a gentle curve with a chasing hammer.

JEWELRY-DESIGN TIP

A curved shape will fit more comfortably on your wrist than will a flat piece.

3 Enamel the piece in 2 layers of Medium Fusing Clear.

4 Dip the top (rounded) side of the bracelet in Sapphire enamel. Fire. Continue to fire layers of enamel on the piece until satisfied.

TORCH-FIRED TIP

When you work with larger metal pieces, you're not able to get parts of the metal as hot as you would when enameling something smaller. How hot the metal is when it comes into contact with the enamel determines how much enamel will stick to the surface. The best advice is to stop counting the layers and use your eye as the guide when enameling larger pieces.

 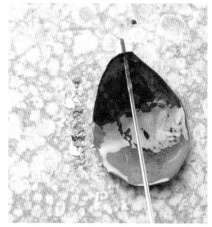 

5 While the enamel is still hot, press the bracelet into a small pile of 6/20 enamel comprised of various colors. Direct the flame to the back side of the piece until the 6/20 fuses to the base enamel.

 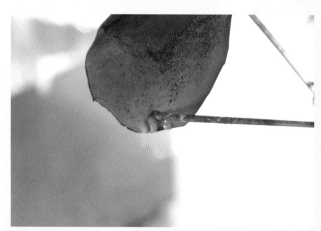

6 Bring the torch topside and lightly fire the 6/20 to round and soften any sharp edges.

7 During the firing, make sure the other bracelet hole remains open. If it starts to fill with enamel, use a mandrel to open the hole.

8 Enamel four 10mm copper discs in Robin's Egg and another 4 in Spruce.

9 Enamel the angel wing in Water Blue enamel. Cool.

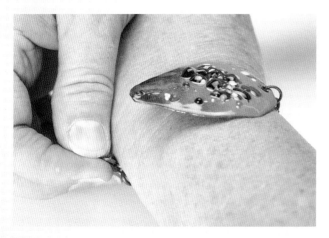

10 Attach 1 end of a chain to the focal piece. With the focal on your wrist, wrap the chain around the remainder of your wrist. Mark that point. Deduct 1" (25mm) for the clasp. Cut the chain.

11 Cut a different style chain to the same length. Begin attaching enameled discs to one of the chains in an alternating color pattern.

12 Use jump ring to join chains. Add the clasp.

13 Attach the other end of the clasp to the bracelet's focal piece.

14 Link several jump rings to create a dangle for the enameled angel wing.

JEWELRY-DESIGN TIP

When the style of the clasp and the design of the bracelet allow, I prefer to have the bracelet open at the side.

MATERIALS

Enameling Tools
basic torch-fired enameling kit

sifter

Enamels
BC303L Clear Liquid Enamel, tinted Bermuda Green

2340 Sea Green

Metalworking Tools
chain-nose pliers

container of water

file, diamond

hole punch

round-nose pliers

scissors

Other Materials and Findings
copper screen, 80 mesh

earwires

head pins, sterling silver, 24-gauge

jump rings, 4mm and 8mm

paintbrush

pearls

sifter

tool for burnishing

TECHNIQUES
Liquid Enamel

Copper Mesh Earrings

I love working with copper mesh. It's lightweight, easily shaped into undulating forms or finely creased folds, and affordable. But it's not perfect. The mesh can produce sharp edges that require smoothing. For right now, though, I'm in love with the pros of the material and will continue to work on the cons. We have a couple of options for getting the gorgeous green of this project. You can either purchase tinted liquid enamel or simply sift Sea Green enamel onto the clear liquid clear enamel while it's still wet.

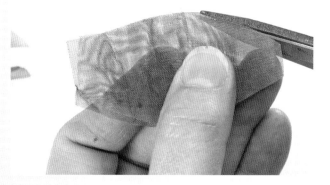

1 Stack 2 layers of copper mesh. Cut out a leaf shape. Each layer will become 1 earring.

2 To shape each leaf, use a pen, mandrel or other object to press a center vein in the leaf. Try not to pierce the mesh.

3 Fold up the edges of the mesh. Use scissors to trim close to the fold. Fold again to create a double fold and to conceal the cut edges of the mesh.

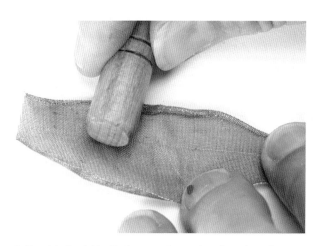

4 Burnish the fold with the cap of a marker, the edge of a pencil or whatever you have.

 JEWELRY-MAKING TIP

When folding the edges, lay the copper mesh on a table. Work on the edge farthest from your body.

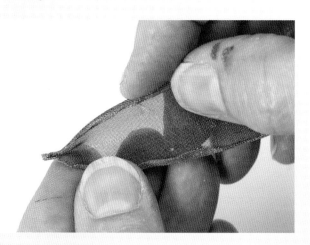

5 Pinch the narrow end of the leaf. Try to tuck in any frayed edges. Grab the end of the leaf and give it a twist.

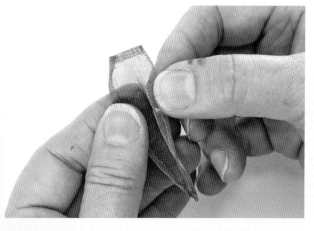

6 Fold over the top edge of the wider end of the mesh.

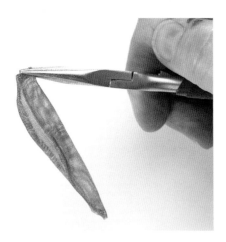

7 Fold the mesh in half and squeeze the top part closed with chain-nose pliers.

8 Punch a hole in the top of the leaf. Repeat steps 2–8 for the second earring.

9 Compare the size and shape of the two leaves. Make necessary changes to create a matching pair. During the firing, you'll also have a chance to tweak the shape with pliers or tweezers.

10 Use a stiff brush to work the tinted liquid enamel into the mesh.

 TORCH-FIRED TIP

If the mesh is particularly dark from oxidation or grime, pickle it before adding enamel.

11 Gently fire the mesh in the flame.

12 Sift Sea Green onto the hot metal and fire. Repeat steps 10-12 for the second mesh leaf.

13 Use 8mm jump rings to attach the leaves to the earwires.

14 Use 24-gauge sterling head pins to wire-wrap pearl dangles onto the earwires.

15 Use 4mm jump rings to attach the pearl dangles to the 8mm jump rings.

JEWELRY-MAKING TIP

When attaching the pearl dangles, remember that you want to creating a mirror image of the first earring.

Hammered

MATERIALS

Enameling Tools
basic torch-fired enameling kit

80 Mesh Enamel
2215 Egg Yellow

2760 Mauve

2840 Mandarin Orange

2880 Woodrow Red

2839 Red

Metalworking Tools
ball-peen hammer, riveting hammer, nail set or center punch

bench block, steel

chain-nose pliers

chasing hammer

chisel, buttonhole cutter or other thin blade

cross-locking pliers

hole punch

pickle

pickle pot

vise

wire cutters

Other Materials and Findings
brass discs, 24-gauge, 1¼" (3cm), annealed, 5

camera, optional

chain, antiqued bronze, 2mm × 3mm, 24" (61cm)

clasp, ball-and-hitch

cording, leather

jump rings, antiqued bronze, 8mm

marker

TECHNIQUES
Firing Flat Objects

Firing Transparent Enamel

Enameling Brass

Annealing

Pickle

Pendant Removal Options

With some inexpensive metal and a little muscle, you can pull off this necklace pretty easily. I love the shine of transparent enamel over these brass discs! Transparent enamel also helps to accentuate any depressions or details in the metal. Textures created by hammer blows, punched holes and dimples are given a boost when the enamel pools in their recesses.

METALWORKING TIPS

- Even though these discs have pre-punched holes for hanging, ignore the holes for now and create new ones when you lay out your composition but before you enamel the discs. The process of forming the hammered discs is the same for all. As you fold and hammer the metal, it will become work hardened and stiff, at which time it will need to be annealed. However, in the end, each will be uniquely different from its neighbor on the necklace.

- Turn on the pickle pot now; you'll need it for step 9.

1 Anneal all discs. Quench in water and dry. Grab opposing sides and fold the disc.

2 Place the fold of 1 disc in a vise and tighten to create a sharp crease.

3 Hammer the disc, if necessary, to further define the crease. Anneal, quench in water and dry.

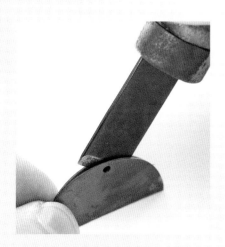

4 Partially open the disc with a thin but not necessarily sharp blade.

5 Put the crease in the vise. Hammer the sides flat against the jaws of the vise. Anneal, quench in water and dry.

6 Pry open the fold. When the fold is open enough so that it becomes a *V* shape, hammer the disc to flatten.

METALWORKING TIP

I use a buttonhole cutter to open my discs, but a 5-in-1 painter's tool is an easy find at the hardware store.

7 Lay the disc with the fold up. Create texture by hammering with a riveting or ball peen hammer, nail set or center punch.

8 Punch decorative holes in the disc, leaving room near the edges for functional holes that will accommodate jump rings. Repeat steps 2–8 for the remaining 4 discs.

9 Place the distressed discs in a pickle pot to remove oxidation from the metal. Rinse and dry all discs.

METALWORKING TIP

The hole punch I used for step 8 is dull. As a result, it pushes out the holes instead of cutting them out; this creates a raised edge around each hole. If your hole punch is sharp and new, you don't have to wait until it becomes dull to get the same effect. Punch your holes. Then simply lay your metal on top of a dapping block—be sure to position a hole over a depression in the dapping block. Tap on the dapping punch with a hammer, thereby raising the sides of the hole.

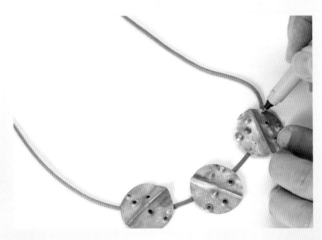

10 Lay a cord in a necklace shape on top of your work surface. Lay the 5 discs on the cord, making note of the direction of the creases and how they will affect the composition. Use a permanent marker to mark the hole placement on the opposing sides of the discs. Take a photograph of the arrangement.

11 Punch the holes at the indicated marks.

12 Enamel each disc in different transparent enamels.

13 Refer to your photo. Link discs together using 3 jump rings between every 2 discs.

14 Measure the length of the linked discs and subtract this number from the desired length of your necklace. Double this number and cut the chain. Pass 1 end of the chain through the jump ring. Use a jump ring at the other end to join both ends of the chain to a clasp. Attach the clasp to the last disc in the necklace.

Water Lily

MATERIALS

Enameling Tools

    basic torch-fired enameling kit

80 Mesh Enamel

    2030 Medium Fusing Clear

    2520 Aqua

6/20 Mesh Enamel

    2030 Medium Fusing Clear

    2410 Copper

    2215 Egg Yellow

Metalworking Tools

    chain-nose pliers

    crimping pliers, optional

    cross-locking pliers or tweezers

    hole punch

    wire cutters

Other Materials and Findings

    bead-stringing wire, 19-strand, .024"

    brass flower, 55mm diameter

    brass flower, 15mm × 17mm

    chain, copper-plated

    clasp, copper-plated

    copper beads, 16mm, 10

    copper disc or copper gear, 25mm

    crimp beads

    earring stop, rubber

    flower head pin

    glue or clear nail polish

    jump rings, 5mm and 7mm

    pearls, 10mm, dyed wasabi green, 13

    twisty tendrils (3)

TECHNIQUES

    Head Pins, Tendrils, Flowers

    Firing Transparent Enamel

    Pendant Removal Options

Simple clear enamel is responsible for the turquoise color of these copper beads. Yes, it's true; the beads are just a conglomeration of turquoise freckles! The secret is in the firing technique. These beads have a mirror finish, similar to that of a Victorian gazing ball.

1 To get the special effects of this firing method, be gentle with your initial heating. Reduce the flame intensity and heat a 16mm copper bead just to the point that enamel will stick to the metal.

2 Apply a layer of Medium Fusing Clear enamel.

SAFETY TIP

These copper beads are heavier than most of the other beads you'll work with. Make sure the mandrel is sized appropriately for the bead hole and that you hold the mandrel parallel to your work surface during firing. Failure to do so may result in the bead sliding down the mandrel and burning your fingers. See *Torch-Fired Tips,* page 83, for more information.

3 Fire the bead. Apply 2 more layers of enamel, for a total of 3 layers.

4 Continue to fire the bead with a low flame. As the dark copper oxides are absorbed in the clear enamel, the bead will become a golden color.

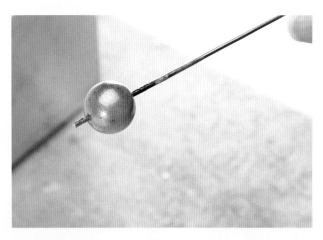

5 Crank up the heat. Overfire the bead until tiny speckles appear on the surface. Be sure to keep the bead in the sweet spot of the flame.

6 Remove the bead from the mandrel. Repeat steps 1–6 for the remaining copper beads.

7 Repeat the firing technique described in steps 1–5 on the large brass flower.

8 As you slowly fire, or *roast* the bead, as my son, David, calls it, you'll see the metal becoming lighter in color. Continue to fire until satisfied. Remove your bead from the mandrel and allow to cool.

TORCH-FIRED TIPS

- Since the copper beads are heavier, they retain heat longer than filigree beads. As a result, allow the enamel to cool a little longer than usual before dropping the bead into the bread pan. This will prevent the bread pan medium (Oil-Dri or vermiculite) from sticking to the enamel.

- During the overfiring process, bright specks of light will appear. These are copper oxides coming from the metal of the bead that, upon cooling, will be turquoise or aqua specks. Sometimes they will turn into little rusty spots, which I think is a nice contrast. This is usually a result, though, of not fully covering the metal with Medium Fusing Clear enamel during the first dredging.

- When metal is in the presence of oxygen, oxides form on the surface. When metal is heated, this oxidation process is accelerated. When copper, brass or sterling silver is heated with a flame, copper oxides are deposited on the surface of the metal. This is what makes the metal look dark when it is heated. Opaque enamels simply cover the dark metal. However, this oxidized metal will significantly affect the fired color of transparent enamel.

- Your goal is to reduce the oxides formed during the heating process and to trap those that do form in clear enamel, known as flux. To reach that goal, lightly heat copper just to the point that enamel will adhere to the surface. Quickly dredge the copper through clear enamel. As you fire copper, the flux prevents the production of more copper oxides and, more important, traps the copper oxides that has already been produced. When you continue to heat copper, copper oxides will be released from the flux. You will gradually see your copper piece becoming golden in color. When it does, apply clear enamel or any number of beautifully colored transparent enamels.

9 Gently heat a small brass flower. Dredge through Medium Fusing Clear. Fire.

10 When the metal has lightened in color and become golden, dredge it through Aqua enamel and fire it.

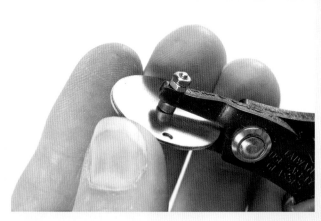

11 Feel free to tweak the shape of the flower during the firing process. When you've achieved the desired result, remove your bead from the mandrel and allow cool.

12 Use a hole punch to place 2 small opposing holes on a copper disc (or buy a disc with holes already in place).

13 To assemble the flower, stack the flower beads as shown onto a glass flower head pin. Add the copper disc.

14 To secure, slide a rubber earring stop onto the head pin wire. Push the components snugly together. Place a drop of glue or clear nail polish at the joint so that the earring stop does not wiggle loose. Wrap the wire.

15 Thread 1 crimp bead onto a 16" (41cm) piece of bead-stringing wire. Bring the wire up through the center of a jump ring, back out, and through the crimp bead.

16 Secure the jump ring to the beading wire by crimping the crimp bead with crimping or chain-nose pliers.

17 Thread on 2 pearls. Then begin alternating pearls and enamel copper beads until your necklace has reached the desired length the desired length. Add 2 more pearls. Close off with a crimp bead and jump ring. (The necklace here uses 13 pearls and 10 16mm copper beads, but you can use more as less as desired.)

18 Use a jump ring to attach 1 end of the beaded necklace to the copper disc.

19 Use a jump ring to attach a length of chain to the other end of the beaded necklace.

20 Attach a copper-plated clasp to the other end of the copper-plated chain. Secure the other part of the clasp to the remaining hole in the disc.

21 Wrap a twisty tendril around the crimp beads at each end of the necklace.

22 Wrap another twisty tendril around the base of the flower so the flower will not flop forward when the necklace is worn.

MATERIALS

Enameling Tools

basic torch-fired enameling kit

Enamels

BC1070 White Liquid Enamel

1055 White Enamel

2030 Medium Fusing Clear

Metalworking Tools

chain-nose pliers

needle file

round-nose pliers

scissors

Other Materials and Findings

copper mesh, 80-mesh

copper wire, 22-gauge

marker

paintbrush

polymer clay bead

round-nose pliers

sifter

steel cable choker

tool for burnishing

tube wringer

TECHNIQUES

Liquid Enamel

Driftwood

Monochromatic color schemes give us the perfect opportunity to focus on form. This project features 80 mesh copper mesh and a tube wringer, like those used by artists to squeeze the last little bit from their paint tubes.

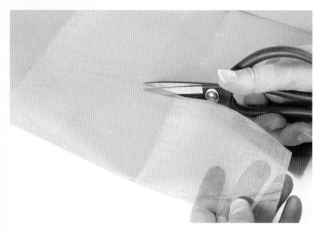

1 Cut a piece of copper mesh to approximately 1½" × 10" (4cm × 25cm). The piece will widen at the center and taper toward the ends. At the widest point it will be 3" (8cm).

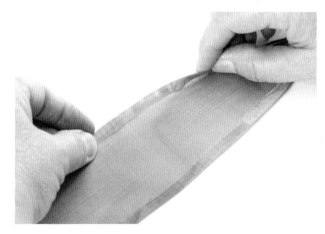

2 Double fold the edges around the circumference of the piece. When folding the edges, it helps to have the copper mesh laying flat on a table. Work on the edge furthest from your body.

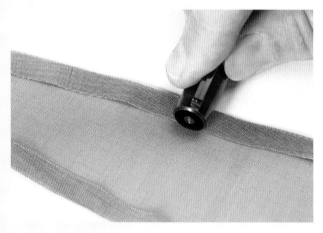

3 Burnish the fold. Here I'm using the cap from a marker, but you can use any item with a hard, round surface.

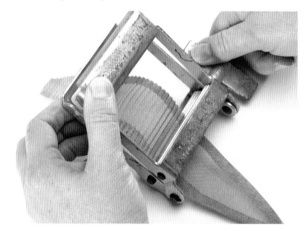

4 Place the narrow end of the mesh in the tube wringer and corrugate the metal.

5 Repeat, except this time slide the mesh into the tube wringer on the diagonal. Re-adjust the placement of the mesh periodically.

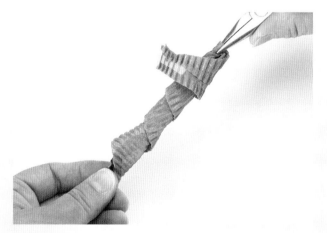

6 Start twisting the loop into a pleasing shape. Remember, you're the designer, so there's no right or wrong way.

For bonus projects and more visit: CreateMixedMedia.com/Mastering-Torch-Fire.

87

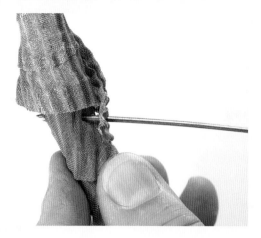

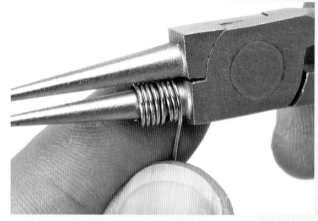

7 Use a needle file or other pointed object to create a hole through the mesh in which to place the mandrel.

8 For a cocoon bead, cut approximately 18" (46cm) of 22-gauge copper wire. Start tightly coiling the wire around the widest part of your round-nose pliers.

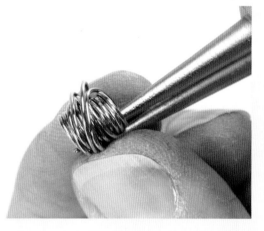

METAL-WORKING TIP

The tight coil creates a channel for the mandrel and for your choker.

9 After you have created a 3/8" (9mm) coiled bead from the wire, release the tension and create looser, more relaxed coils around the tight coil.

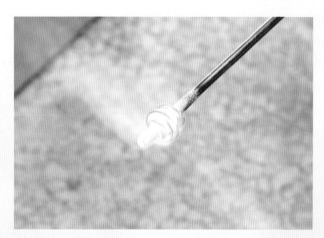

10 Enamel the cocoon bead in 1 layer of Medium Fusing Clear and 2–3 layers of White. Remove the bead from the mandrel and allow it to cool.

11 Insert the mandrel into the hole in the mesh. Dip the corrugated mesh piece into the liquid enamel. (You may brush the liquid enamel onto the piece, if you prefer—use a stiff paintbrush.)

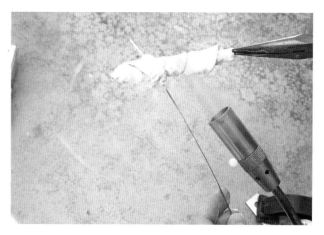

12 Remove the saturated mesh piece from the liquid enamel and allow excess liquid enamel to drain off.

13 Grab the other end of the mesh piece with pliers to help stabilize the piece and to more easily move the piece through the flame. Fire the piece.

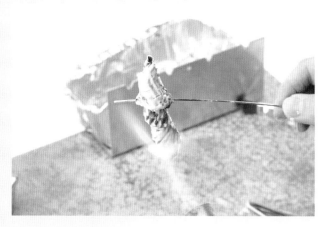

14 Liquid enamel does not get shiny like powdered enamels, so look for the metal to glow orange as a sign that the enamel is maturing.

15 If added coverage is needed, sift on White enamel. Fire. When you are satisfied with the bead, remove it from the mandrel and allow it to cool.

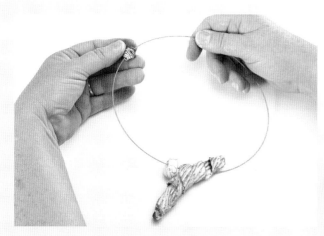

16 Thread both elements and a polymer clay bead onto a stainless choker. These chokers come with a magnet permanently attached to one end. You'll glue on the magnet at for other end after you've finished sliding on the elements.

Gizmo

MATERIALS

Enameling Tools

basic torch-fired enameling kit

80 Mesh Enamel

1710 Tallow

1715 Clover

1760 Iris

2030 Medium Fusing Clear

6/20 Mesh Enamel

1239 Mellow Yellow

1410 Robin's Egg

Metalworking Tools

brass mallet

cross-locking pliers or tweezers

dapping block and punch

disc cutter

Eugenia Chan three-hole punch

riveting tool

round-nose pliers

X-ACTO knife

Other Materials and Findings

bottle cap

copper disc, ¾" (2cm)

copper disc, 1" (25mm)

copper eyelets, 1.8mm

copper gear, open spoke, ¾" (2cm)

dapping block and punches

glue, quick-set

leather lacing, 12" (30cm)

marker

scrap leather

sewing machine or leather needle, optional

thread, upholstery-weight

wire, copper, 22-gauge wire

TECHNIQUES

6/20 Enamel

Firing Flat Objects

Head Pins, Tendrils Flowers

Annealing

Pendant Removal Options

Leather is hot right now! As the price of metals has increased, the interest in leather as a jewelry-making element has also increased. The two may be related, or it might simply be that people love leather! It's not necessary to buy special leatherworking tools—for this project you'll use the hole punch and riveting tool you've seen used in the other projects. Stop by a thrift store and repurpose an old leather belt, jacket or skirt, or find what you need in the scrap bin at your favorite craft store.

1 Use a 1" (25mm) gear with a center hole as a guide to punch a hole in a 1" (25mm) disc. Punch $^{3}/_{32}$" (2mm) holes on opposite sides of the disc (or purchase a disc with holes already in place, like I did!).

METALWORKING TIP

When deciding what size hole to punch in your metal, make it larger than you need because enamel will narrow the hole.

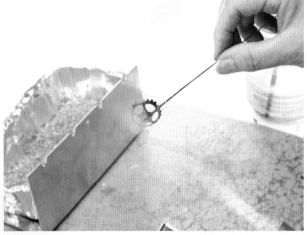

2 Use a ¾" (2cm) gear with a center hole as a guide to punch a center hole in the ¾" (2cm) disc. Do not punch holes in the sides.

3 Anneal the ¾" (2cm) disc and gear, then quench them and let them dry.

4 Dap the ¾" (2cm) gear and disc using a 38mm dapping punch.

5 Enamel the now domed ¾" (2cm) gear in Tallow enamel. Remove the gear from the mandrel and allow it to cool.

6 Enamel the 1" (25mm) disc in Iris. Use the Bead Pulling Station to push the disc away from the end of the mandrel by about 2" (5cm) while keeping the mandrel parallel with the tabletop. Allow the disc to cool on the mandrel.

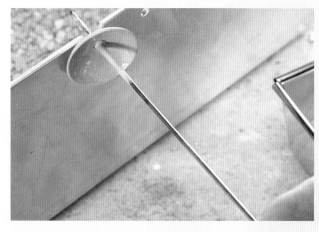

7 Place the domed ¾" (2cm) disc on a mandrel with the convex side facing away from your hand. Enamel the domed ¾"(3cm) disc in Clover. Use the BPS to push the disc from the end of the mandrel by about 2"(5cm). Allow the disc to cool on the mandrel.

8 Add the gear you enameled in step 5 to the mandrel with the domed disc. Heat.

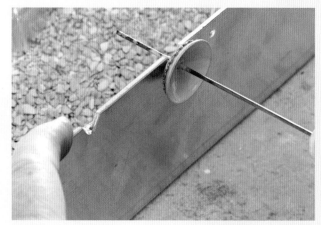

9 Use the Bead Pulling Station to push the 2 pieces together. Make sure that the gear is centered on the dapped disc. Remove this compound piece from the mandrel and let it cool.

10 Place the compound piece you created in steps 8 and 9 on the mandrel that holds the disc you created in step 6. Heat the pieces and join them by pressing them together with the Bead Pulling Station.

11 Sift Medium Fusing Clear enamel onto the piece. It will act as an additional glue to hold the pieces together. Fire the piece.

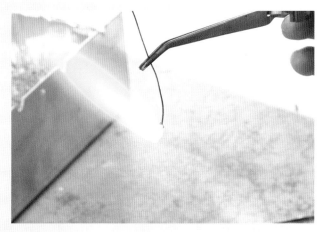

12 Cut two 4" (5cm) pieces of 22-gauge wire. Hold the midpoint of the wire with cross-locking tweezers. Enamel a glass head on each end using 6/20 enamel in Mellow Yellow and Robin's Egg.

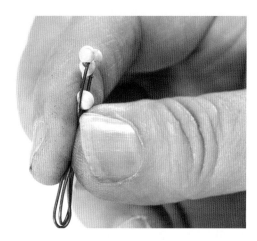

13 Fold the wires up so they create a bundle.

14 Carefully wrap the wires around the tips of your round-nose pliers to create twisty tendrils.

15 Secure the bundle by adding a drop of strong, fast-setting glue to the fold.

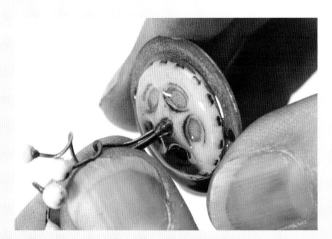

16 Push the bundle through the hole in the top of the enamel piece. Add more glue, if necessary.

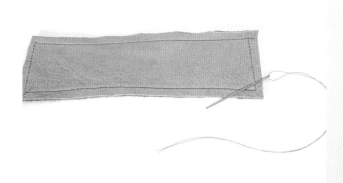

17 Balance your enamel creation on a bottle cap while the glue dries.

18 Cut a piece of scrap leather measuring 6" × 1¾" (15cm × 4cm). If you cut your own piece of leather, prevent it from stretching by stitching a border ¼" (6mm) from the outside edge using upholstery thread. If you use a sewing machine, use a leather needle and a Teflon or walking foot. Alternatively you could also punch holes around the edge with a hand punch and then sew a running stitch through the punched holes.

19 Determine the center of the leather band, either by eye or by measuring it. Place the compound enamel piece on the leather and mark the rivet placement with a marker.

20 Use a disc cutter to create two small ¼" (6mm) discs. You can also cut out small squares of metal if you don't have a disc cutter. Punch a ³⁄₃₂" (2mm) hole in the center of each disc. These pieces will prevent the eyelets from pulling through the leather.

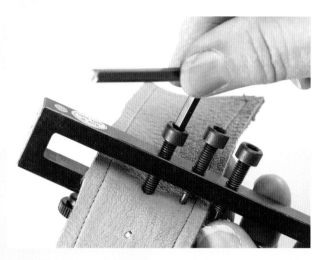

21 Punch ³⁄₃₂" (2mm) holes in leather.

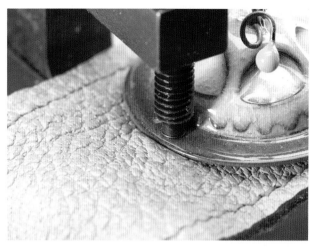

22 Thread an eyelet through a hole in the compound piece, the topside of the leather strap and the 6mm disc. Use a riveting tool to compress and flare the eyelet. Repeat for the other side of the compound piece.

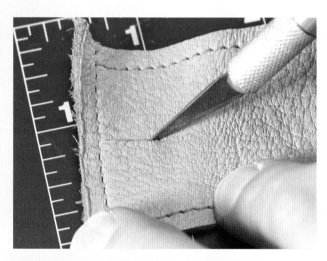

23 With an X-ACTO knife or buttonhole cutter, create a slit at each end of the leather strap.

LEATHERWORKING TIP

To prevent the knife from slipping and cutting too large a hole, stick a piece of tape on the leather and mark the slit placement and size. Lift one end of the tape and place a straight pin at the end of the mark for the slit. Repeat for the other end. The pins will prevent you from accidentally making too large an opening.

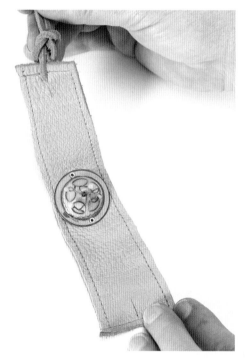

24 Tie leather lacing through 1 of the slits.

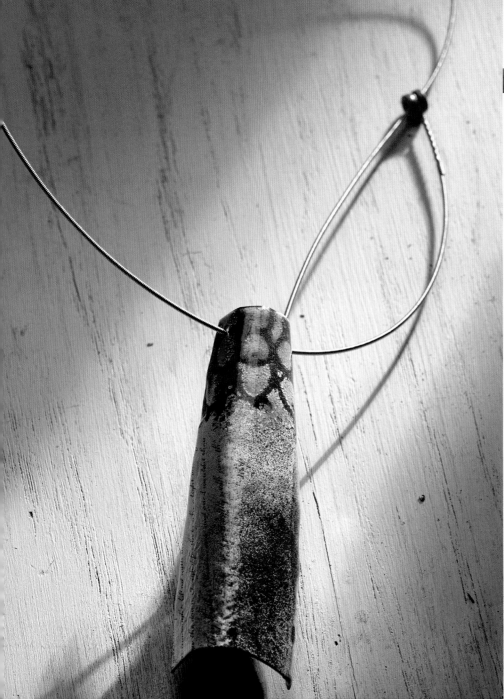

The Great Composer

When I designed this necklace, I knew I wanted to make something with a more architectural influence. I thought about ordering steel wire from a jewelry supplier and then I looked across the hall. There, surrounded by musical instruments, was my son, David. "David, do you have a broken bass string?" I asked. "Actually, I just broke one last night," David replied. I didn't know bass strings had a handy ring at one end of the string that would become instrumental in guiding the path of the choker. One of the serendipities of the necklace design is that the length is adjustable; the excess bass string is easily tucked away behind the pendant when a shorter length is desired. This necklace features fire-scale decoration, a ceramic decal and liquid enamel.

1 Cut a shape from copper: approximately 1" (25mm) (at the top) × 1½" (4cm) (at the bottom) × 3" (8cm) (long). Anneal, quench, and dry. Lay the metal on the rubber bench block. Hammer the center of the piece lengthwise with a chasing hammer. The side of the metal will cup upwards.

2 Use pliers or a the flat side of a chasing hammer to coax the metal into shape.

3 Pickle the metal to remove the oxidation created by annealing.

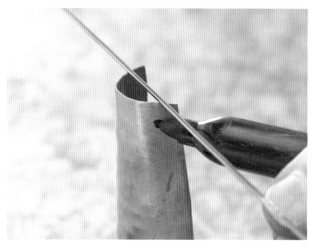

4 Stand the piece on end. Lay a mandrel across the narrow end of the pendant. Visualize the path of the choker. Mark the holes directly under the mandrel on each side. Punch a ⅛" (3mm) hole at each mark.

 METALWORKING TIP

Punch a larger hole than usual. Remember, enamel will narrow the hole. We also want to allow for some wiggle room for the choker.

5 Paint white liquid enamel onto the copper or, for more even coverage, dip your copper piece into liquid enamel. Let the enamel dry (a couple of minutes should do it).

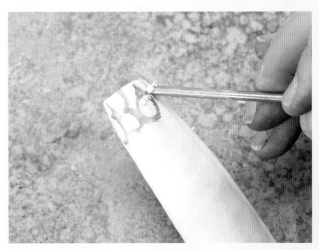

6 Scratch a design through the dried white liquid enamel with a mandrel.

☀ TORCH-FIRED TIP

The scratching through of 1 layer of enamel to reveal another surface is called *sgraffito*. Allow the liquid enamel to dry to the point that it is still damp. (Don't worry, handling it will not leave fingerprints.) There is an increased tendency for the liquid enamel to chip off when creating your design if the enamel is completely dry.

7 Fire the liquid enamel, and then dredge the piece through Nile Green enamel. Fire.

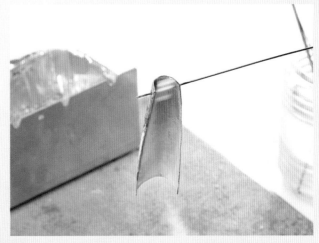

8 Add another layer of Nile Green, if you'd like. Fire. Allow the pendant to cool.

TORCH-FIRED TIP

Oxidation, for the most part, can be a nuisance during many phases of jewelry creation, but it can be especially friendly to an enamel artist. The *Great Composer* calls for pickle, a mild acid that removes oxidation from metal. When you apply enamel to cold metal as you do when working with liquid enamel, the metal must be clean of oxidation, oil from fingers, dirt and grime from environmental exposure.

The sgraffito technique creates a design by removing some of the white liquid enamel, which in turn reveals bare copper. When the liquid enamel is dry and the project is fired, the white liquid enamel remains white, but the exposed copper becomes dark brown. This is a result of the oxidation process, which is accelerated when metal is heated.

A couple of layers of Nile Green over the piece will create a beautiful aqua color over the white liquid enamel but will not affect the dark brown color of the sgraffito design.

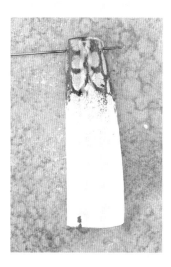

9 Print a decal with a handwritten font. Tear off a piece of the decal for the pendant. Soak the decal in water for about 30–45 seconds.

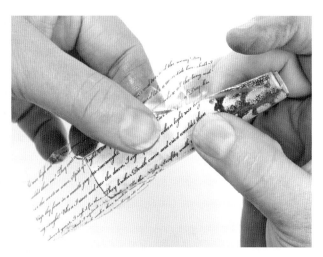

10 Slide the decal from its paper backing onto the white liquid enamel of the pendant.

11 Trim any excess decal from the edges of the pendant.

12 Blot water from the decal.

13 Refire the piece, heating it gradually to prevent thermal shock and the build up of steam behind the decal. Steam can cause a *blowout* of some of the decal pattern.

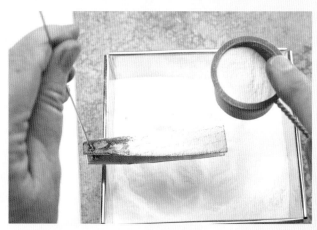

14 Sift Medium Fusing Clear enamel over the decal to protect it.

TORCH-FIRED TIP

Be careful not to overfire the decal because the image could burn out completely or become so light in appearance as to not be appreciated. When possible, avoid direct flame contact with the decal.

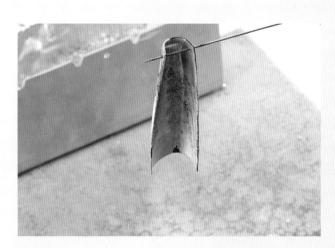

15 Fire the pendant and allow it to cool.

16 Add water to Cornflower Blue liquid enamel. Apply a painterly brushstroke to 1 side of the pendant. Fire.

17 When the pendant is a glowing orange, sift Orient Red enamel down the center of it and fire. Use the Bead Pulling Station to remove the piece from the mandrel and cool.

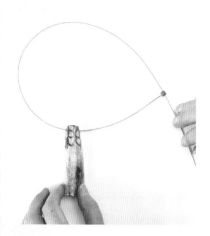

18 Thread a G bass guitar string through the two holes in the top of the pendant.

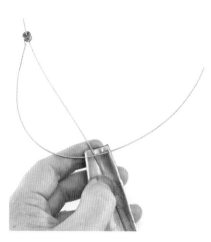

19 Thread the bass string behind the choker part of the necklace.

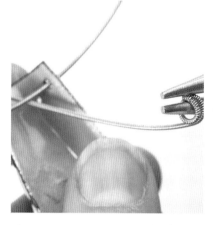

20 Curl the loose end of the guitar string with round-nose pliers.

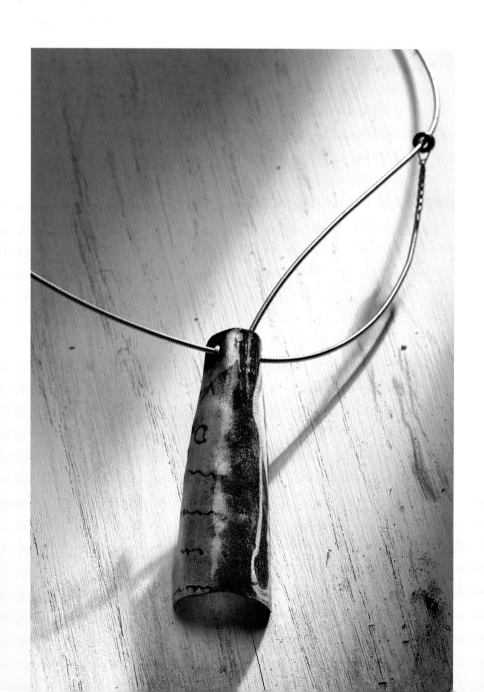

Forest Nymph

Can you imagine woodland nymphs frolicking by a stream with the lucky one wearing this piece of perpetual beauty? Time to get your head out of the clouds and back to basics in the studio where we'll explore the fundamental qualities of metal: how to work-harden it and then make it pliable again. A few simple techniques will allow you to create organic shapes with ease. Glass flower head pins, whether formed by an expensive press or with low-tech tools, like tweezers, are more than fillers in the design. Juicy transparent glass is contrasted with enamel flowers with darkened edges.

MATERIALS

Enameling Tools

basic torch-fired enameling kit

medium sifter

bell flower glass press

80 Mesh Enamels

1055 White

1319 Bitter

1410 Robin's Egg

1415 Sea Foam

1720 Mauve

1830 Marigold

1870 Orient Red

6/20 Mesh Enamels

1239 Mellow Yellow

1830 Marigold

2410 Copper

2715 Rose Purple

2836 Red

Metalworking Tools

ball-peen hammer

beading tweezers

bench blocks, rubber and steel

chain-nose pliers, 2

chasing hammer

Crock-Pot

cross-locking pliers

dapping block and punches

Eugenia Chan three-hole punch, flexible shaft drill or Dremel tool

marker

metal shears

round-nose pliers

vermiculite

wire cutters

Other Materials and Findings

beads, assorted seed beads, spacer beads, etc.

chain

clasp

copper sheet, 30-gauge

copper wire, 14-gauge

copper wire, 24-gauge

jump rings, 8mm

ruler

variety of twisty tendrils and glass bell flower head pins

TECHNIQUES

6/20 Enamel

Firing Flat Objects

Head Pins, Tendrils, Flowers

Firing Transparent Enamel

Annealing

Pendant Removal Options

FORMING THE NECKPIECE

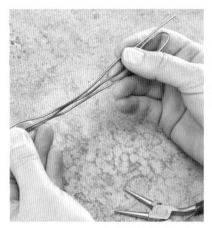

1 Cut 24" (61cm) of 14-gauge copper wire. Mark the wire 8¼" (21cm) from each end.

2 Fold the wire at each mark to create 3 parallel wires. The outside wires should extend slightly beyond the middle wire.

3 Lay the parallel wires on top of a steel bench block. Begin at 1 end, and hammer the entire length of the wires. The wires will curl upward, creating a neckpiece that will rest comfortably on the collarbone.

METALWORKING TIP

Hammering wire on a steel bench block will flatten, thin and stiffen the wire, which will help it maintain its shape.

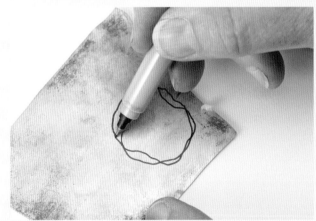

4 Hold the neckpiece so the curve is facing you. With your palm facing up, grab the end of 1 of the cut wires with round-nose pliers. Create a loop by rotating your hand toward you. Bring the end of the wire through the parallel wires. Repeat for the other side. Set it aside.

5 Use a marker to draw an undulating circle on the metal.

MAKING THE FLOWERS

Create the flowers for the necklace from varying sizes of 30-gauge copper squares, from ½"–2½" (13mm–63mm). Because enamel adds weight and strength to the metal, a 30-gauge sheet is the perfect thickness for creating delicate flowers.

TORCH-FIRED TIP

- Make accommodations for the fact that forging and enameling will narrow the hole in the metal. Be safe and make the hole larger than you would normally.
- Forging work hardens the metal and makes it resistant to shaping. Be prepared to anneal the metal more than once during the forming process to make it more pliable.

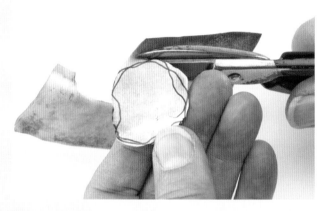

6 As you cut out the shape with metal shears, rotate the metal, not the shears.

7 Using the Eugenia Chan three-hole punch, punch a ³⁄₃₂" (2mm) hole near the center of the piece.

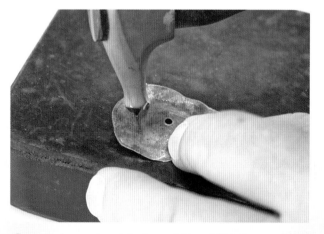

8 Place the metal flower on a rubber bench block and hammer the center with a chasing hammer.

9 Anneal, quench and dry the metal. Lay it face down on the rubber bench block. Hammer in from the edge with a ball-peen hammer. Encourage the edges to turn upward by lifting an edge of the flower and hammering on the opposite side.

TORCH-FIRED TIP

The Eugenia Chan three-hole punch is designed to reach the center point of the varying sizes of metal we're using in this project. The other alternative would be to use a flexible shaft drill or Dremel tool.

10 Using a dapping block, dap the flower for a greater center depression.

11 Tweak and curl the edges of the flower with round-nose pliers to create an organic shape. Repeat steps 1–11 to make several flowers (or petals) of various sizes, which together will comprise the focal point of the necklace.

12 Now is the time for some planning. Because the shape of the flower cannot be changed after enameling, all forging and shaping must be completed before enameling can begin. Pair your flowers and examine how well they nest.

13 To improve fit, stack the flowers on a rubber bench block. Hammer the center of the nested flowers with a ball-peen hammer or dapping punch. The petals may need an adjustment, too.

14 Enamel the smallest flower with the color of your choosing and make the edge a different color. For the remaining flowers, fire each flower in 2 to 4 layers of enamel (in colors of your choosing). Larger flowers will need more layers. Roll the edges of some of the flowers through other enamel colors. Add other enamel colors to the centers of smaller flowers. You can do this by pinching enamel between your finger and thumb, and dropping it on the flower center.

15 When satisfied with the color and firing results, remove the flowers from the mandrel.

⚡ TORCH-FIRED TIP

This is a very colorful project! Feel free to use any enamels you have on hand. Chances are they're some of your favorite colors and you can't go wrong!

16 Stack the flowers onto an enameled bell flower head pin.

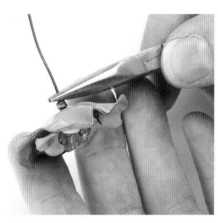

17 Add a 3mm bead onto the wire as a spacer.

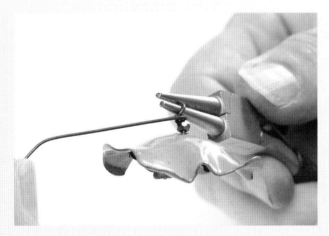

18 With round-nose pliers, create a wire-wrapped loop. Trim any excess wire.

19 Open a jump ring and slide the jump ring through the loop.

ASSEMBLING THE NECKLACE

20 Cut 24" (61cm) of 24-gauge copper wire. To anchor 1 end of the wire, weave it through the parallel wires of the neckpiece. String a few beads on the wire and continue weaving until you have approximately 3" (8cm) of woven beading. Use small beads at places where you make turns and change the direction of the wire.

21 Continue adding beads and weaving wire, and stop when you have reached a spot where you would like to add an enameled flower.

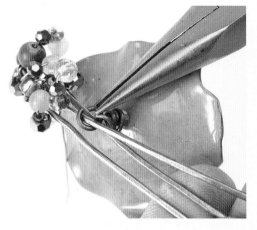

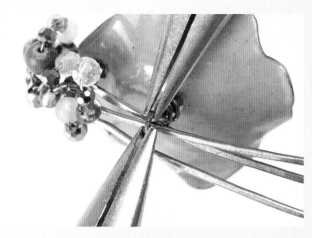

22 Slide the jump ring around a strand of the neckpiece.

23 Close the jump ring.

 JEWELRY-MAKING TIP
You can directly wire wrap flowers to the neck-piece. I prefer to add them with jump rings because doing so allows me an opportunity to more easily change their placement.

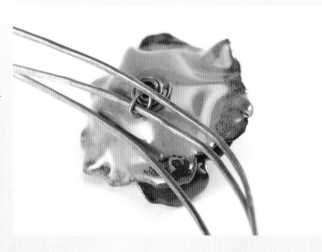

24 Continue weaving beaded wire into the neckpiece to secure the flowers in place. Add twisty tendrils and bell flowers to hide exposed wires.

25 Attach 1 end of the chain to the neckpiece with a jump ring. Attach a clasp to the other end of the chain with a jump ring.

INNOCENCE

For a downloadable PDF with instructions for making **Innocence**, visit createmixedmedia.com/mastering-torch-fire.

Flow

One of the special things about enamel is that a metal pendant can become your canvas. I have a special attraction to loosely drawn circles and lines that can add structure to a design. This project also features instructions on how to create a mold of a twig and then to use that mold to create a pewter casting. So take a walk in the yard, find that special twig and get started!

MATERIALS

Enameling Tools
basic torch-fired enameling kit

Enamels
1202 Off-White

1410 Robin's Egg

1860 Flame Orange

2115 Mars Brown

Green Overglaze Enamel

Lilac Overglaze Enamel

P-1 Overglaze Enamel

2030 Medium Fusing Clear

Metalworking Tools
bench block, rubber

chain-nose pliers, 2

chasing hammer

cross-locking pliers or tweezers

hole punch

metal shears

wire cutters

Other Materials and Findings
butane torch

chain, snake

copper sheet, 24-gauge

eyedropper

flux

frit tray or aluminum can

jump ring

liquid dish soap (such as Dawn)

molding compound

paintbrushes

patina solution (such as Nova-can)

plastic spoon

small containers, 2

sanding pad

solder, lead-free, soft

soldering iron

twig

wax paper

wire, copper, 16-gauge

TECHNIQUES
Overglaze

Firing Flat Objects

Pendant Removal Options

1 To create a mold of a twig, follow the mixing directions for the molding compound. Roll the molding compound into a log shape. Press the twig halfway into the molding material. Allow the mold to cure. To test: Press your fingernail into an inconspicuous part of the mold. If your nail leaves a mark, allow more curing time. Remove the twig from the cured mold.

2 Place a coil of lead-free soft solder onto a frit tray or in an aluminum can.

 TORCH-FIRED TIP

I was unable to use an aluminum can because the ones I tried were lined with plastic. After 2 instances of setting off the smoke alarm, I remembered these cool *frit trays* from Arrow Springs. Frit is composed of small chips of glass. Lampworkers and others working in hot glass roll molten glass through frit. The frit tray I use has a convenient pouring spout and handle.

3 Holding the tray in your dominant hand and the butane torch in your nondominant hand, heat the solder until molten.

4 Pour the molten solder into your mold and allow it to cool completely.

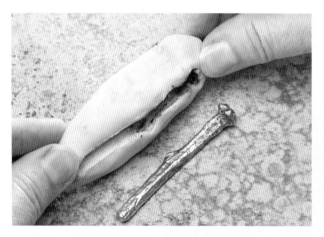

5 Pop the cast twig out of the mold.

6 Cut 10" (25cm) of 16-gauge copper wire and clean it with a sanding pad. Bend the wire in half to create a bail. Flux the wire and the twig with flux for soft soldering. Solder the bail to the back of the pewter twig. Allow the wire ends to extend approximately 3" (8cm) beyond the twig. Clean excess flux off the twig using a liquid dish soap that has no moisturizers.

7 Use Novacan patina solution to age the twig and the copper wire of the bail.

8 Use metal shears to cut an irregular oval, approximately 1¾" × 1½" (45mm × 13mm), from a copper sheet. Lay the sheet metal on your work area and your bail apparatus on top. Properly align the pieces. Come down about ⅜" (9mm) from the top edge of the pendant and mark the placement of 2 holes for hanging the pendant.

9 Punch 2 holes, each ⅛", in the top of the pendant.

10 Anneal, quench and dry the pendant.

11 Lay the pendant on a rubber bench block and hammer in the center of the metal with a chasing hammer. The metal will cup upward. Stop when you're satisfied with the shape.

12 Enamel the pendant in 4–5 layers of Off-White. Remove from the mandrel and cool.

13 Place a very small amount of Lilac overglaze enamel (china paint) in a container. Use an eyedropper to dilute with mineral spirits, and mix.

14 Hold the pendant in your hand and tilt it slightly over wax paper. Load a liner brush with the diluted Lilac overglaze enamel and wipe the brush along a narrow section of the upper edge of the pendant. Allow the diluted overglaze to flow in a loose line down the front of the pendant. Repeat with a diluted Green overglaze enamel (china paint).

15 Pinch Robin's Egg enamel in tweezers and lay it on some of the wet overglaze.

16 Repeat step 15 with 1 layer of Flame Orange enamel.

17 Refire (follow the directions on *Refiring Enamel* on page 25).

18 Dredge the pendant through Medium Fusing Clear enamel and heat until glowing. Clear enamel will give the pendant a glassy surface. Remove the pendant from the mandrel.

19 Load a liner brush with P-1 overglaze enamel. Paint your design. If necessary, thin the P-1 with mineral spirits to improve brushability.

20 Refire to fuse the overglaze enamel (follow directions for *Refiring Enamel* on page 25). Remove the pendant from the mandrel. File the edges, if necessary.

21 From the front, thread the bail wires through the holes.

22 When the twig is positioned as desired, bend the wire ends up.

23 Wrap the wires.

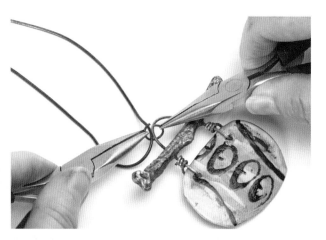

24 Attach the pendant to a chain with a jump ring.

BOUND

Bound, a variation of the project **Flow**, is quite easy to re-create by simply following the enameling directions set forth for **Flow**. The bail, however, combines a rectangular block of Woolly Mammoth Ivory, horsehair and leather. To create the bail, simply drill holes in the ivory that corresponds to the location of the pendant holes. Attach the ivory to the pendant and create the bail. Lash horsehair onto the ivory, and finally wrap leather over the bundle of horsehair and around the back side of the ivory. Do some crisscrossing of sterling wire to bind everything together and you're finished!

Reef

How could I name this necklace anything but **Reef**? **Reef** makes me want to dive into a pool of inviting bubbles (but only if the water is 95°F)! A little soldering is required in this piece, but I promise—it's a cinch. The secret to **Reef's** uniqueness is the enameling technique. Isn't that why you're here? With clear enamel and a special firing technique, we'll turn copper to gold and then put on the glitz! So let's dive in!

MATERIALS

Enameling Tools
basic torch-fired enameling kit

80 Mesh Enamel
2030 Medium Fusing Clear

6/20 Enamel
2305 Nile Green

2410 Copper

Metalworking Tools
bench block, steel

chain-nose pliers

chasing hammer

hole punch

needle file

riveting hammer

round-nose pliers

scissors

wire cutters

Other Materials and Findings
baking soda

bead stopper

beading needle

butane torch

C-Lon cording, aqua

clamshells, sterling silver, 2

clasp, magnetic

container of water

copper bead, oval, 1¼" × 1" (3cm × 25mm), 1

copper beads, round, ⅞" (2cm), 3

crimp bead

firebrick or other heat-resistant surface

flux brush

flux for hard soldering

glue

jump rings, sterling, 5mm and 10mm

liquid dish soap

liver of sulfur (LOS)

marker

pickle

pickle pot

sanding pad

scrubbing pad

silver beads, 3mm faceted

solder, sheet

tweezers or soldering pick

wire, sterling, 16-gauge

TECHNIQUES

6/20 Enamel

Firing Transparent Enamel

Annealing

Soldering

Pickle

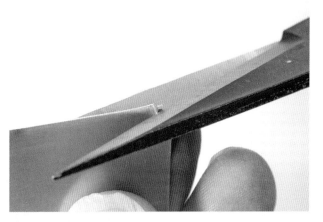

1 Cut 4½" (11cm) of 16-gauge sterling wire. Use semiflush cutters to recut the ends of the wire. At each end, make sure the flat back of the cutters is toward the length of wire. A little point will be raised on each wire end. Use a needle file to create a flush surface.

2 Clean the sheet solder and ends of the wire with a sanding pad to remove any traces of oxidation or grime. Cut a 2mm × 2mm square from the sheet solder. This is called a *pallion*.

METALWORKING TIPS

In this project you're going to do *hard soldering* with silver solder. This is different from *soft soldering* which uses materials purchased from the hardware store. Silver soldering can be a finicky process. It requires that all of the elements, including the solder, the flux and the metal being soldered, be extremely clean or the solder won't flow. It also requires that the ends of the wires to be soldered meet precisely because solder does not fill gaps. Hold your joint up to the light. If light passes through the join, re-file the joint until you can see no light. If the solder balls up but does not flow after continued heating, put the piece in a pickle (a mild acid) and start over.

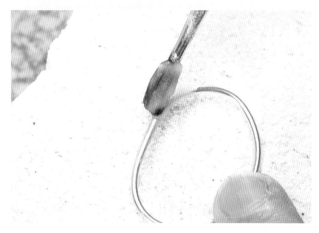

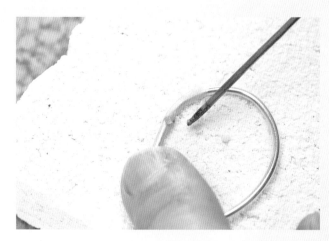

3 Lay the wire ring on a firebrick (or other soldering surface). Paint the joint with a small amount of flux for hard soldering.

4 With tweezers, a flux brush or a soldering pick, place the pallion of solder on the joint.

5 Circle the wire slowly with the flame of a butane torch. The wire will be heated during this process, and the water in the flux will evaporate. When the flux becomes clear and shiny, direct the flame to the seam until the solder flows. Remove the flame immediately.

6 Dip the ring in water and wash with a scrubbing pad and Dawn dish soap to remove any residual flux.

7 Hammer the ring with a chasing hammer to flatten it. Then, use a riveting hammer to create texture. Use a marker to indicate hole placement. Punch a hole on opposing sides. Smooth any rough areas with a file.

8 Create a solution of liver of sulfur (LOS) by placing a LOS nugget in warm water.

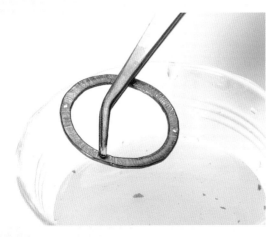

9 Place the ring in the solution.

10 When satisfied with the patina, neutralize the LOS by placing it in a solution of 1 teaspoon of baking soda to a half cup of water to stop further oxidation. Rinse in clear water, and dry.

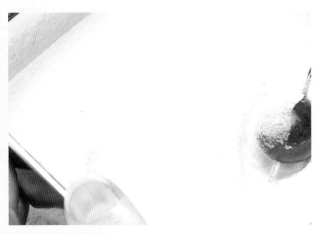

11 To enamel a bead, lightly heat it to the point that it just begins to glow. Don't overheat! Quickly and completely dredge the entire bead in Medium Fusing Clear.

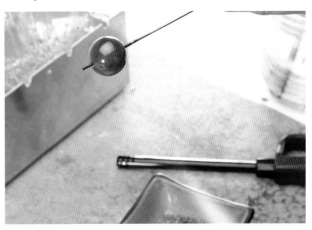

12 Gently heat the bead until the copper oxides that have been trapped in the clear enamel begin to be absorbed by the clear enamel. As this happens, the bead will become golden in color.

13 While the bead is still hot and on the mandrel, lay it on a steel bench block and forge it into an organic shape with a chasing hammer.

14 As the bead cools, reheat it and forge it some more.

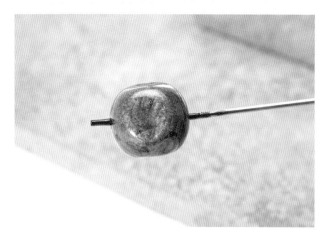

15 When satisfied with the shape of the bead, add another layer of Medium Fusing Clear and begin to overfire the bead by turning it slowly in the flame. Copper oxides in turquoise shades will begin to surface from the copper and tint the clear enamel.

16 When the bead is glowing hot, roll it in a mixture of 6/20 Copper and Nile Green enamels.

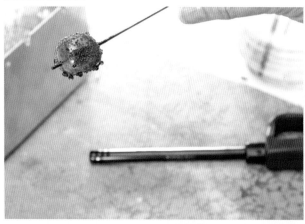

17 Be gentle with the flame and watch for the 6/20 to melt. It should look like rounded beads of glass with no sharp edges. When you are sure the 6/20 has adequately fused to the base enamel, remove from the bead from the mandrel to cool. Repeat steps 12–17 for the remaining 3 beads.

18 Cut 18" (46cm) of C-Lon cording and connect it to a 5mm jump ring with a Lark's head knot. Use a beading needle or create one from a folded fine gauge wire.

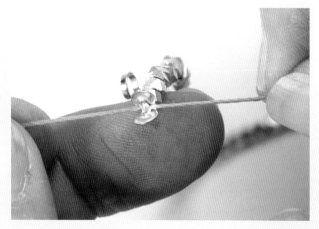

19 Thread on another 45 more faceted silver beads. Thread on a clamshell, knot the thread.

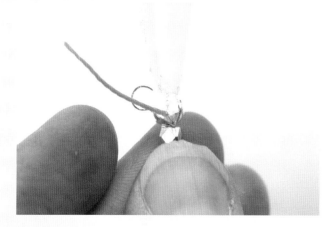

20 Add a drop of glue to the knot.

21 Squeeze the clamshell with pliers to crimp.

22 Cut two 10" (26cm) lengths of C-Lon. Place a 5mm jump ring at 1 end.
String on 7 faceted silver beads, 1 enameled copper bead, 7 faceted silver beads, 1 enameled copper bead, 7 faceted silver beads, 1 enameled copper bead and 25 faceted silver beads. Thread on a crimp bead.

Pass the cord through a 5mm jump ring and back through the crimp bead. Crimp the bead with flat-nose pliers.

Thread any excess C-Lon through several of the silver beads.

23 Slide a 5mm jump ring through the 5mm jump ring at the end of each strand. Slide the new 5mm jump rings through the holes in the silver ring, and close them.

24 Thread a 10mm jump ring through the loop in the clamshell and attach a magnetic clasp. Repeat for the other end of the necklace.

Bandaged Heart

Even without a hydraulic press to dome your pieces, you can use the low-tech alternative detailed here to achieve a similar look. Actually the effects created by hand-forging are quite special. Every tap of the hammer contributes to a piece that can never be duplicated, not even by you! The pendant detaches from the necklace, and thereby allows you to make many beaded necklaces to go with this one pendant. Wear the heart with a simple chain, enamel links, a beaded necklace; my mind is racing with ideas! In this project, we'll be painting with Black Overglaze Enamel, sawing and hard and soft soldering.

MATERIALS

Enameling Tools

basic torch-fired enameling kit

Enamels

1055 White

2115 Mars Brown

P-1 Black Overglaze Enamel

BC1070 White Liquid Enamel

Metalworking Tools

bench block, rubber

bench pin

chasing hammer

dapping punch

drill bit

eyedropper

files, metal

household drill, flexible shaft drill or Dremel tool

jeweler's saw with spiral saw blade

metal shears

wire cutters

Other Materials and Findings

butane torch

chain, copper

copper bead, 16mm

copper cleaner (such as Penny Brite)

copper screen, 80 mesh

copper sheet, 24-gauge

duct tape

enameled bead, 14mm

firebrick or soldering block

flux brush

flux for hard soldering

flux for soft soldering

head pin

jump ring, 8mm

leather cording, brown, 2mm

marker

paintbrushes

plexiglass, 4" × 4" × ¼" (10cm × 10cm ×6mm)

pickle

pickle pot

sanding pad, fine

scrubbing pad

solder, soft

solder, silver, hard

T-pins, steel

tool for burnishing

wax paper or paper plate

wire , copper,16-gauge

TECHNIQUES

Liquid Enamel

Firing Delicate Shapes

Annealing

Soldering Pickle

1 Draw a simple heart shape directly on the plexiglass paper. Leave at least ⅜" (9mm) all the way around your drawing.

2 With a household drill, flexible shaft or Dremel tool, drill a pilot hole for the spiral saw blade. Drill just inside the line of the drawing.

JEWELER'S SAW PRIMER

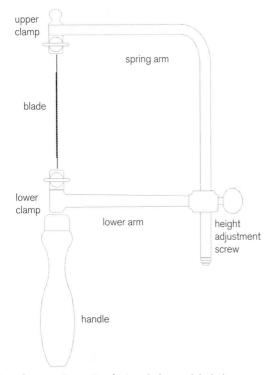

upper clamp

spring arm

blade

lower clamp

lower arm

height adjustment screw

handle

For reference, the parts of a jeweler's saw, labeled.

Spiral saw blades are designed to cut plastic and wood multi-directionally, unlike a common jeweler's saw blade which cuts in one direction.

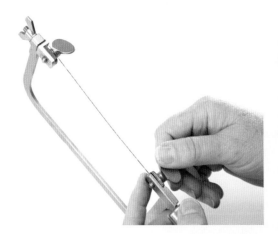

Assembling a Jeweler's Saw, Part 1
Place 1 end of the spiral blade in between the 2 pieces of metal at the lower clamp. Tighten the wing nut.

Assembling a Jeweler's Saw, Part 2
Rest the top of the spring arm in a bench pin and the handle against your sternum. Lean into the saw frame. Align the saw blade between the 2 metal squares at the upper clamp. Tighten the wing nut. Pluck the blade and listen for a high pitch to indicate that the tension is sufficient to proceed.

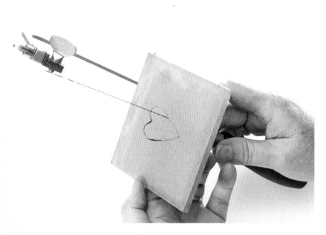

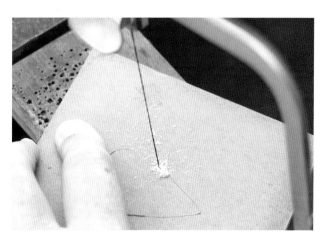

1 Attach the blade at the lower clamp. Thread the saw blade through the hole in the plexiglass. Slide the plexiglass template so that it rests against the lower clamp. Rest the spring arm against a workbench or table and the handle against your sternum. Lean into the saw frame. Align the saw blade between the 2 metal squares at the upper clamp. Tighten the wing nut. Pluck the blade and listen for a high pitch to indicate that the tension is sufficient to proceed.

2 Rest the plexiglass on your bench block. Saw on the line of your shape, or as close to the line as possible. Loosen 1 end of the saw blade to remove the blade from the interior space of the template.

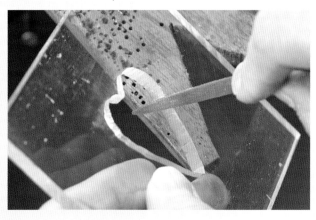

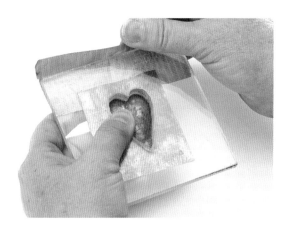

3 Remove the paper from plexiglass. Refine the shape with a metal file, if necessary. Anneal the copper.

4 Place the annealed square of copper roughly the cutout in the plexiglass. Duct-tape the metal to the plexiglass.

5 Create a depression by tapping on the metal with a chasing hammer.

6 A dapping punch is also useful to define the outer edges of the heart shape.

7 Hammer the top of the metal near the cutout to create a defined shape.

8 Remove 1 end of the tape to check your work. Look for flattened areas in the shape that need more doming. If more work is necessary, replace the tape and continue. If satisfied, remove the copper from the plexiglass.

9 Cut out the heart. Accentuate the 2 lobes of the heart by removing a small wedge of metal. If the shape needs further defining, place it in the plexiglass form and hammer.

10 File the edges of the heart.

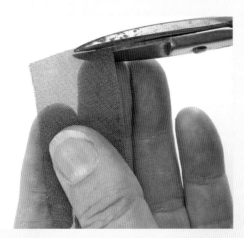

11 Cut a strip of 80-mesh copper screen approximately 1" × 2" (25mm × 5cm).

12 Clean the mesh by placing it in pickle for 2 minutes.

13 Fold the mesh in half lengthwise.

14 Reinforce the seam by burnishing it.

15 Fold back the edges of the copper, leaving a standing seam in the center of the piece.

16 Place a pea-sized amount of White Liquid Enamel (dry form) on a piece of wax paper or other disposable surface. With an eyedropper, place a few drops of water on the enamel. Mix.

TORCH-FIRING TIP

When mixing dry form liquid enamels, be stingy with the water; very little is needed.

17 Brush the White Liquid Enamel onto the mesh, making sure you get enamel in between the strands of mesh.

18 Hold the mesh with a pair of cross-locking tweezers and fire in the flame until glowing. The mesh can be delicate. Be careful with the flame or you'll end up with a hole (which could be a great addition to the design!). Let the mesh cool completely.

19 Clean the heart in order for the enamel to fuse. Place it in a pickle for a couple of minutes or scrub it with a copper cleaner (like Penny Brite). Rinse in water.

20 Bend the enamel mesh to take on the form of the heart. The slight cracking noise you hear is OK. The enamel cracks will heal when you refire the piece.

21 Trim the mesh piece so it fits on the heart.

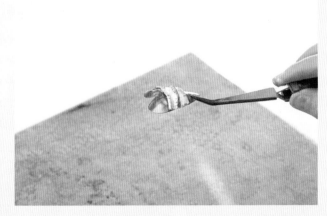

22 Paint liquid enamel onto the back of the mesh and place it on the heart. This will act as a temporary adhesive.

23 To attach the mesh to the heart, hold the heart with pliers and direct the torch flame to the underside of the heart. Press the mesh onto the heart to secure the connection.

24 Cut approximately 2" (5cm) of 16-gauge copper wire. Curve the wire with your hands to create a bail.

25 Hammer the ends of the wire to flatten and thin them.

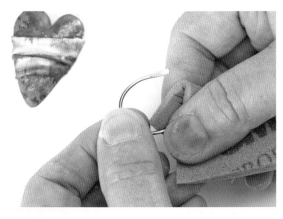

26 Use a fine sanding pad to clean the parts of the metal heart and ends of the wire that will be soldered. If necessary, pickle the heart again to ensure it is clean enough to take solder.

27 Lay the heart face down on a firebrick soldering block (or other surface for soldering). Position the bail on the heart. Make sure the wire ends and the metal of the heart are in direct contact. Use props, such as a stainless steel T-pin or steel wire, to prop up the bail. Use a paintbrush to apply a small amount of flux for hard soldering to the back of the heart and to the ends of the wire.

28 Clean a piece of hard sheet solder with an abrasive like a sanding pad. Cut into the side of the solder like fringe. Cut off two pieces of fringe (pallions). Each piece should be approximately 2mm × 2mm.

29 Pick up the pallions with a flux brush and use the brush to place the pallions so they are touching the ends of the bail.

For bonus projects and more visit: CreateMixedMedia.com/Mastering-Torch-Fire.

129

30 Heat the copper, wire, solder and flux with a butane torch until the solder melts. After the solder melts, it will flow to the area where the metal is hottest. Heat the areas around the wire, and the solder should flow to those areas and envelop the wire, thereby attaching it to the heart.

31 Clean the bail with an abrasive (sandpaper, scouring powder, etc.). Rinse and dry the bail. Cut ¼" of soft solder from the coil. Flux the solder and the bail.

METALWORKING TIP

The Hot Head is too robust for the soft solder and will scorch it, so for this type of soldering use a butane torch. If you've never tried this technique, you might want to practice a little bit first.

32 Heat the solder with a butane torch until it melts and forms a ball. Immediately place the bail in the middle of the ball of solder until the solder envelops the wire. If there are rough spots, heat those spots lightly with the butane torch until the solder flows.

33 Slightly flatten the solder with a hammer.

34 Cut a piece of leather cording approximately 40" (102cm) long. Fold the cording in half. Find the center of doubled cording and move the center point about 2" (5cm) toward the folded end of the cording. This is necessary because you will be tying 2 knots—instead of 1—at the cut end.

Fold the leather cording in half at the new center point. Pass the center point through the bail, from the front to the back. Bring the cut ends of the cord through the resulting leather loop and create a Lark's head knot.

35 Create an overhand knot. Push the knot down toward the bail and Lark's head knot to keep the knots compact.

 TIP

This project requires more leather cording than you might normally expect, because you have to allow extra length for the knots.

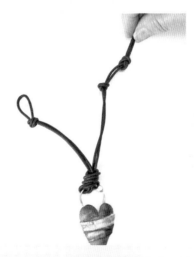

36 Tie another overhead knot in each end of leather cord.

37 Tie another knot in the cord (closer to the heart) that accommodates the 16mm enamel bead.

38 Cut the chain to the desired length. Double it . Pass the chain through the loop end of the leather cord that is not knotted. Attach the other end of the chain to an enamel bead.

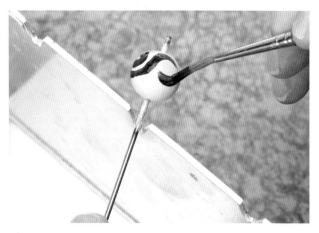

39 Paint a swirl design on the white pre-enameled bead with P-1 Overglaze Black Enamel. Let the overglaze dry.

40 Fire the bead with 2 layers of Mars Brown. Allow the bead to cool.

41 Create a bead dangle. On a head pin, thread on a 3mm heishi bead, a 4mm brass heishi bead, the enamel bead, a 4mm brass heishi bead and a 3mm heishi bead. Create a simple wire-wrapped loop in the head pin.

42 Catch both ends of the chain with a jump ring, attach the bead dangle and close the jump ring.

43 Slide the bead through the knotted loop to secure the necklace.

Contributor Gallery

How can we not think about color when we think of enamel? Color is the true essence of enamel. But color, or at least our way of working with it, can become too familiar and predictable. We return to Old Reliable—our palette that has become safe because it has served us well. But sometimes we surprise ourselves when we stretch and exercise our imagination to incorporate what we might consider an oddball or offbeat color into a composition only to find that it's **that** color that brings the piece to life! Many times color inspirations for a project will come from an incongruous assortment of beads where adjacent beads will spark a flight of thought! And the excitement begins …

Enamel artists Lynnea Bennett, Laura Guenther, Heather Marston and Carol Meyers chose to accept an assignment of working with color schemes based on the color wheel. The color wheel can inspire your creations. But if you're still a little timid and need more time for ideas to percolate, look to textiles for color inspiration. I find that doing so is one of the most sure-fire ways to come up with a winning color combination. Use this inspiration while you train your eye to see what works and what doesn't. Be an observer.

The color wheel can be very useful in helping us to understand color harmonies when we design our jewelry.

Complementary Colors: In jewelry design, compositions that feature complementary colors are some of the most dynamic, but the compositions can sometimes be jarring. This is where a tone or tint of a hue paired with a saturated version of the hue could create a lively composition that is not too jarring. See the jewelry submission by Lynnea Bennett.

Monochromatic Colors: When we select one of the "wedges of pie" on the color wheel and begin at the outside ring and work toward the center, we create a monochromatic color scheme. These arrangements are usually quiet, calm and allow other features of the work, such as texture and shape, to stand out. See the jewelry submission by Laura Guenther

Analogous Color: Analogous colors are those next to each other on the color wheel and usually consist of one primary color. They present a rich composition that is somewhat monochromatic and calming. See the jewelry submission by Carol Myers

Split Complementary Colors: Split complementary colors are found when you select a color but instead of selecting the color directly opposite on the color wheel, which would be the complementary color, you select the colors to each of the sides of the complementary color. Split complementary color schemes create energetic compositions; ones that you might initially think are risky, but in fact they're usually just beautiful! See the jewelry submission by Heather Marston.

JEWELRY-DESIGN TIP

If complementary and split complementary color schemes seem a little intimidating, remember that you don't have to select the saturated version of each hue for your composition. Consider using a saturated color for one of the hues and a tone or tint for the other(s). Refer to the books of Beverly Ash Gilbert (.gilbertdesigns.net) for inspirational jewelry compositions that get their direction from the color wheel.

The Color Wheel:
The Way I See It in Thompson Enamels

Monochromatic Color Scheme

Grandpa's Journey
By Laura Guenther
PWF Certified Instructor
Greensboro, NC

blueantiquities.blogspot.com
blueantiquities.Etsy.com

An old photo of a moment in time long passed took me back to family summer trips to my grandparents' home in South Dakota. With a cheek pressed against the glass of the car window, I would gaze at the never-ending farmland dotted with white houses, red barns and slow-moving cows on our way to Mitchell, South Dakota. We knew when we arrived a treat would be waiting for us, usually a trip with Grandpa to buy root beer for root beer floats.

My Grandpa Carlson was the inspiration for my necklace with a monochromatic color scheme. I created this design using several torch-fired enameling techniques, including torch-fired ceramic decals, beads that were rolled through enamel and the details of impressed metal accentuated by liquid enamel.

Opaque enamels used in the piece include 1345 Hunter Green, 1315 Willow Green, 1308 Lichen, and 1335 Pea Green. I also used 2340 Glass Green. I blended the enamels on the medium-size beads by enameling the entire bead in one enamel, dipping the bead midway in a second enamel and rolling the circumference of the bead in a third enamel. For the floral connector, I used liquid enamel and scraped away the raised areas before firing. Long days and hard work on Grandpa

Carlson's farm were just part of his life's journey, which is represented in the enameled photo connector and a decal photo similar to a photo I have of my grandparents standing outside their farm.

My goal in creating this necklace was to capture the essence and the spirit of an earlier time when hard work on the farm was celebrated by root beer floats with family.

Split-Complementary Color Scheme

Confetti

By Heather Marston
Leola, PA

hehebeads.wix.com/cswdesigns
cswdesignsbyhehe.blogspot.com/
hehebeads.etsy.com
facebook.com/HeHeBeads
Twitter: @CSWDesigns

It took some trial and error to find a split complementary color scheme that inspired me but after getting yellow, green, orange and violet involved, I knew I was on to something.

The enamels used in this project are:

- Yellow-orange—two layers 1830 Marigold, one layer 2215 Egg Yellow, 6/20 enamel in 1319 Bitter
- Yellow-green—two layers 1319 Bitter, one layer 2230 Lime Yellow
- Violet—two layers 1760 Iris, one layer 2720 Harold and one layer 2660 Nitric

Once the colors were selected, the design came next. I laid out my metal pieces and moved them around several times before finally coming up with a preliminary design. With my pieces laid out, I selected the enamel colors for each piece. The focal disc shows dots of 6/20 enamel in 1319 Bitter. I really love the combination of yellow orange and yellow green. The violet color of the large copper gear provides just the pop the piece needed.

A brass hexagon was selected as part of a toggle clasp. The textured metal was a bit too shiny for the piece so I used alcohol inks to color the brass to match the violet gear. The toggle part of the clasp features balled-up wire on both ends, and the ends are enameled in 6/20 enamel in 1319 Bitter. 6/20 enamel

is similar in size to small pea gravel or Kosher salt and creates a great head pin.

I admit I started the process with uncertainty. I felt pushed out of my comfort zone and I had to work with a color scheme that isn't part of my regular palette. That said, I really love the way this necklace turned out. The split complementary color scheme was a challenge but I enjoyed every step of the process.

Analogous Color Scheme

Blue Bonnets

By Carol Myers
PWF Certified Instructor
Frederick, MD

facebook.com/caroldeezigns
caroldeezigns.blogspot.com

Since many of my favorite colors adjoin each other on the color wheel, from a pale sage green to vibrant turquoise, I was thrilled to create a project that represents an analogous color scheme of harmonious colors. I spent a full day firing the colors I adore and another developing the composition for this project.

I created beads with subtle hue shifts by layering different transparent enamels over opaques. Many of the beads also showcase one of my favorite effects, which is the blending of color when an opaque enamel bead is rolled through a transparent enamel. This process randomly layers the transparent enamels, creating additional colors.

The necklace includes a fan of textured sheet metal created by my daughter, Stevie Ballow. Not only is the fan an integral part of the design, but working with her on this project made the experience all the more special!

The opaque enamels used in this project are: 1055 White, 1308 Lichen, 1319 Bitter, 1422 Aquamarine, 1920 Stump Gray, and 1940 Steel Gray.

The transparent enamels used are 2230 Lime Yellow, 2305 Nile Green, 2310 Peppermint, 2520 Aqua, 2660 Nitric Blue.

Complementary Color Scheme

Falling Leaves

By Lynnea Bennett
Cincinnati, OH

designsbylynnea.com
facebook.com/Designsbylynnea
flickr.com/photos/40372277@N03/
designsbylynnea.blogspot.com/
twitter.com/DesignsbyLynnea

Every year my workbench is full of colorful leaves I've enameled. This year was no exception except that these metal leaves are vintage beads purchased during a visit to a New York City specialty supplier. I convert these enamel pieces into necklaces and earrings. This complementary color project relies on transparent enamels to alter and boost the color.

The colors were created by using the following enamels:

- Gold/Yellow—two coats of Lemon (1225) with one coat of Egg Yellow (2210)
- Blue Violet—two coats of Fox Glove (1745) and one coat of Nitric (2660)

I used a brass chain and wire wrapping, and finished this project with a recycled sari ribbon.

Index

About Barbara Lewis

This section might be the hardest part of the whole book to write! How do you sum up a life that has spanned 63 years? I guess what's of most importance is that I love to learn and I love to teach others what I learn. My BFA with a concentration in ceramics and nearly twenty years working as a ceramic artist was the springboard to Painting with Fire Studio (PWF). I simply applied the principles of firing a 40 cu. ft. gas kiln to firing a small gas torch.

PWF, a family-owned business, initially focused on travel teaching and an online store that kept students supplied with the materials of their new favorite addiction! The naming of *Torch-Fired Enamel Jewelry: a Workshop in Painting with Fire* as Best Craft Book of 2011 at Amazon propelled PWF into a brick and mortar store/studio. In June 2012, Painting with Fire Studio opened in midtown St. Petersburg, Florida, with a state-of-the-art flameworking learning center. Currently, PWF has embarked on the Painting with Fire Teacher Certification Program, which will allow the Immersion Process of torch-fired enameling to be taught by qualified teachers throughout the U.S. and in other countries.

Finally, as I write this, we're in the process of establishing a metal arts learning center, a natural extension of my participation in an 18-month bench jeweler's program. It is my good fortune to live blocks from the only public technical school in Florida with such a program.

My love of learning continues to influence my direction and destiny. We strive to make PWF Studio a cooperative learning space. We teach you what we know and gladly learn from others who are willing to share their knowledge and experience. The many good people who have taken an interest in our success have blessed us beyond measure.

Resources

- Painting with Fire Studio
  2428 Central Avenue
  St. Petersburg, FL 33712
  (727) 498-6409
  www.paintingwithfirestudio.com
  Patented bead pulling station,
  Thompson Enamels, enameling
  and metalworking tools, Eugenia
  Chan tools, custom-tinted liquid
  enamels

- Rio Grande
  www.riogrande.com
  sterling silver supplies gemstones, etc.

- Tandy Leather
  www.tandyleatherfactory.com
  leather and leather supplies

- Crafted Findings
  http://www.craftedfindings.com
  riveting and other tools

- Ventilation Design Tutorial by
  Whit Slemmons for Andrea
  Guarino-Slemmons
  andreaguarino.com/
  VENTILATION.html

- Patsy Croft's Enamel and Goldsmith Blog – A great resource!
  alohilanidesigns.com

Dedication

This book is dedicated to my family: Jim, David, Laura and Matt, and the Painting with Fire team of Rachel Meyer and Linda Healy.

Acknowledgments

I would like to thank:

The North Light Team, who consistently produces the best books in the craft market. A special thanks goes to Kristy Conlin, my very talented editor; to Christine Polomsky, photographer extraordinaire; and to Ric Deliantoni for his insightful work on my recent North Light DVDs.

Students, online friends, customers, Ning members and PWF teachers who continue to be a source of encouragement to me and the program.

My many teachers, including my early ceramics teacher, Phyllis Handal, who cut off part of the tail of her Arabian horse for two of my projects. I knew she was nice, I just didn't realize that nobody does this. Her generosity in this, and in many other ways, will not be forgotten.

Joseph Spencer, who taught me the immersion method of enameling.

To Savanna Do, my jewelry professor, who is teaching me the necessity of possessing "high hand skills" when working with precious metals.

To Carrie Boucher, the talent behind Pink Crow Studio (etsy.com/shop/PinkCrowStudio) and the brainchild behind Nomad Art Bus (nomadartbus.org) who seeks to bring "art to all." Carrie generously shares her vast artistic knowledge with me every time we're together.

To Diane Shelly, executive director of the Florida Craftsmen Gallery, who embraced me and Painting with Fire as part of the St. Petersburg arts community and continues to be a support.

To my loving grandmother, who taught me how to first express my creativity at her treadle sewing machine when I was 12 years old.

Metric Conversion Chart

To convert	to	multiply by
Inches	Centimeters	2.54
Centimeters	Inches	0.4
Feet	Centimeters	30.5
Centimeters	Feet	0.03
Yards	Meters	0.9
Meters	Yards	1.1

Mastering Torch-Fired Enamel Jewelry Copyright © 2014 by Barbara Lewis. Manufactured in China. All rights reserved. No part of this book may be reproduced in any form or by any electronic or mechanical means including information storage and retrieval systems without permission in writing from the publisher, except by a reviewer who may quote brief passages in a review. Published by North Light Books, an imprint of F+W Media, Inc., 10151 Carver Road, Suite 200, Blue Ash, Ohio, 45242. (800) 289-0963. First Edition.

Other fine North Light Books are available from your favorite bookstore, art supply store or online supplier. Visit our website at fwmedia.com.

18 17 16 15 14 5 4 3 2 1

DISTRIBUTED IN CANADA BY FRASER DIRECT
100 Armstrong Avenue
Georgetown, ON, Canada L7G 5S4
Tel: (905) 877-4411

DISTRIBUTED IN THE U.K. AND EUROPE BY F&W MEDIA INTERNATIONAL LTD
Brunel House, Forde Close, Newton Abbot, TQ12 4PU, UK
Tel: (+44) 1626 323200, Fax: (+44) 1626 323319
Email: enquiries@fwmedia.com

DISTRIBUTED IN AUSTRALIA BY CAPRICORN LINK
P.O. Box 704, S. Windsor NSW, 2756 Australia
Tel: (02) 4560-1600; Fax: (02) 4577 5288
Email: books@capricornlink.com.au

ISBN 13: 978-1-4403-1174-1

Edited by Kristy Conlin
Photography by Christine Polomsky and Al Parrish
Designed by Geoff Raker
Production coordinated by Jennifer Bass

Ideas. Instruction. Inspiration.

Receive FREE downloadable bonus projects, techniques and more!
Visit CreateMixedMedia.com/mastering-torch-fire.

torch-fired enamel jewelry

a workshop in
painting *with* fire

BARBARA LEWIS

creative
torch-fired enamel
techniques

a BRACELET
workshop

BARBARA
LEWIS

NORTH LIGHT DVD | an artistsnetwork.tv production

COLLAGE
VILLAGE

BIRD ORNAMENTS
PRINTED & STITCHED   p. 67

cloth paper
scissors   COLLAGE   ARTISTIC
             MIXED MEDIA   DISCOVERY

NOVEMBER/DECEMBER 2012   ISSUE 45

light up
YOUR ART
mixed-media style

GET *warm*
and cozy
WITH felting
books, collage,
tools, reader challenge,
and more!

mix it
MATCH IT
4 quick jewelry
techniques
p. 54

Find the latest issues of
Cloth Paper Scissors
on newsstands,
or visit artistsnetwork.com.